Atomic Absorption Spectroscopy

Atomic Absorption Spectroscopy

SECOND EDITION
Revised and Expanded

JAMES W. ROBINSON

Department of Chemistry
Louisiana State University
Baton Rouge, Louisiana

MARCEL DEKKER, INC.   New York

MARCEL DEKKER, INC.
270 Madison Avenue, New York, New York 10016

LIBRARY OF CONGRESS CATALOG CARD NUMBER:
73-00000   ISBN: 0-8247-6249-5

Current printing (last digit):
10 9 8 7 6 5 4 3 2 1

Printed in the United States of America

To My Left Hand, Still Trying

Preface

In the years since the writing of the first edition, atomic absorption spectroscopy has enjoyed a meteoric rise in acceptance and importance in analytical chemistry. It is used routinely in many fields of endeavor from medicine to agriculture. The flame is the established atomizer and its attraction and shortcomings are well documented.

The association with the flame is so strong that all other atomizers are called nonflame. New horizons in analytical chemistry are appearing thanks to such atomizers. Sensitivities of 10^{-14} g are readily obtained for some elements but contamination, storage, interference, and calibration problems are difficult to handle. When we have solved these problems, we will have a technique capable of opening many new doors.

This edition is an attempt to document the state of the art in 1975. It is intended for the practicing analyst, the student, and research worker alike. It will not completely satisfy any one of them but it should satisfy the needs of many.

James W. Robinson

CONTENTS

Atomic Absorption Spectroscopy

Chapter 1

INTRODUCTION

1.1 WHAT IS ATOMIC ABSORPTION SPECTROSCOPY?

Atomic absorption spectroscopy involves the study and measurement of radiant energy by free atoms. The data obtained by studying this absorption provide spectroscopic information and analytical information. The spectroscopic information includes the measurement of atomic energy levels, the determination of oscillator strengths, the population of atoms in various energy levels, atomic lifetimes, and so on. The analytical information revealed includes qualitative and quantitative determination of elements, particularly the metallic elements of the periodic table.

It is the objective of this book to study atomic absorption spectroscopy principally as an analytical tool. To do this, it is necessary to understand both analytical chemistry and the spectroscopy involved in order to generate reliable analytical data.

The analytical process involves the conversion of molecules or ions into free atoms and then the measurement of absorption of radiation by these free atoms. The conversion of solutions of chemical compounds to free atoms involves areas of physics, chemistry, and voodoo—the latter still playing the most important role even after almost two decades of study.

The absorption of energy by atoms follows well-known physical laws and appears to be predictable, thus providing us with a basis for quantitative analytical chemistry. The radiant energy absorbed by the atoms is generally in the form of very narrow absorption lines with wavelengths in the visible or ultraviolet region of the radiant energy spectrum. During the absorption process the outer valence electrons of the atoms jump to a higher orbital and the atom is said to become excited. To a first approximation there is a very simple relationship between excited and unexcited atoms, and therefore between atomic absorption and atomic emmission spectroscopy. This simple relationship is illustrated in Fig. 1.1, which illustrates the equilibrium between an excited atom and an unexcited atom plus a photon. The unexcited atom is said to be in the ground state.

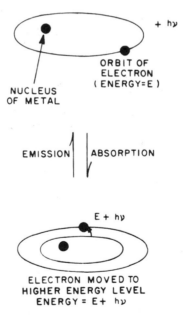

Figure   1.1.   The relationship between atomic-absorption and atomic-emission spectroscopy.

The generation of a ground-state atom plus a photon from an excited atom is the basis of emission spectroscopy. In this technique we measure the number of photons generated in the emission process. The reverse process of absorbing a photon by a ground-state atom to become an excited atom is the basis of atomic absorption spectroscopy. Except under special circumstances we do not measure the "number of photons emitted or absorbed"; rather we measure the intensity of light emitted in the case of emission, or the percentage of light absorbed in the case of absorption.

In any atom there are numerous permitted energy levels and numerous transitions permitted between different excited energy levels and between excited energy levels and the ground state. This will be discussed in more detail later. The existence of numerous upper excited states points out that the equilibrium between ground-state atoms and excited atoms is not as quite as simple as is represented by Fig. 1.1. However, for the sake of clarity and for the purpose of illustrating the principles of emission and absorption, the upper energy states will be ignored at this point, and we shall limit our discussion to the ground state and the first excited state of the atom.

The illustration in Fig. 1.1 helps to reveal the gross effects of variable conditions of a system of excited and unexcited atoms. For example, any change that increases the total population of atoms generated from a given sample will increase both the number of excited atoms and the number of unexcited atoms. This in turn leads to an increase in both the emission signal and the absorption signal. The use of highly combustible and volatile solvents, high atomizer temperatures, and organometallic compounds all tend to increase the number of free atoms formed in a flame atomizer from a given sample concentration. Under these circumstances we can expect that both the emission signal and the absorption signal will be increased. Because there are numerous excited states available to the atom, however, the increase in emission intensity from any particular level does not bear a simple relationship to the increase in atomization effeciency. But with increased atomization efficiency, the number of free atoms increases and the degree of absorption by these atoms increases.

In contrast to these circumstances, if a particular condition favors the formation of *either* ground-state *or* excited atoms (but not both), then we shall get an increase of *either* emission and a drop in absorption or vice versa. An example of this is an increase in flame temperature. For a given population of free atoms, if the atomizer

temperature is increased, there will be an increase in the number of
excited atoms and a corresponding decrease in the number of unexcited
atoms. Thus, we can see that increasing atomizer temperature will
increase both the efficiency of atomization and therefore emission
and absorption; but it will also change the balance of the number of
atoms and produce more excited atoms. It can be seen, therefore,
that the effect of temperature is not the same for emission spectrography
as for atomic absorption spectroscopy. The same can be said for other
variables, and we shall try to understand these effects by considering
each of the variables in turn and how they effect emission and absorption.

One of the most basic and most important relationships in
spectroscopy is given in Eq. (1.1),

$$E = h\nu \tag{1.1}$$

where

E = energy difference between two energy levels in an atom
(or molecule) between which transition occurs
h = Planck's constant
ν = frequency of radiation

Also,

$$\nu = \frac{c}{\lambda} \tag{1.2}$$

where

c = speed of light
λ = wavelength of radiation

The importance of (Eq. 1.1) gives the frequency (and hence the
wavelength) of radiation associated with an energy shift from one
energy state to another in an atom or molecular system. We shall confine
our arguments to atomic energy levels inasmuch as we are principally
concerned with atomic spectroscopy.

The equation tells us that the energy change E in the atom must
equel the energy $h\nu$ of the associated photon.

In emission spectrography an excited atom loses energy and drops
to a lower energy state. In the process it emits a photon with energy equal
to its change of excitation energy.

In atomic absorption spectroscopy, if an atom changes from a ground state to a higher excited state, its energy increases. This energy E comes from the absorption of a photon (energy $h\nu$), whose energy is equal to the energy required to cause excitation in the atom. As an illustration, if an atom has two energy levels of E_x and E_y electron volts, respectively, then the difference between these energy levels equals $E_x - E_y$ electron volts.

If the atom is in the higher energy state, then it will emit a photon with energy equal to $E_x - E_y$ electron volts. If it is in the lower energy level, then it may absorb a photon of energy $E_x - E_y$ electron volts to become excited.

The quantum theory tells us that only certain energy levels are permitted in atoms (or molecules). This means that the atom can exist for extended periods only at permitted energy levels; all other energy levels are unstable and can be ignored in this branch of spectroscopy.

Since only certain energy levels are permitted, the energy difference between these levels is well defined and thus only certain wavelengths of radiation can be emitted or absorbed by an atom. Consequently, the emission spectrum of an atom is characteristic of a particular element, and similarly only certain radiation frequencies can be absorbed by any particular element. These two properties provide us with the specificity necessary to make atomic absorption and emission spectrography a useful, reproducible analytical tool. If an atom such as sodium is found to absorb at a wavelength of 5890 Å, then we can be assured that it will always absorb at 5890 Å because the energy levels associated with the absorption of this wavelength of energy are a physical property of sodium.

If a system of atoms is held at an elevated temperature, a number of the atoms will exist in the upper excited states. The intensity of emission depends on the number of atoms in these excited states. This relationship is governed by the equation

$$S = \frac{N_2 E}{\tau} = \frac{N_1 E}{\tau} \frac{g_1}{g_2} e^{-E/kT} \tag{1.3}$$

where

S = intensity of emission line
N_2 = number of excited atoms
N_1 = number of unexcited atoms
τ = lifetime for the excited atom

E   =   energy of excitation ($E = E_x - E_y$)

$\dfrac{g_1}{g_2}$ =   probability of the ratio of atoms in the ground state and the excited state

T   =   temperature

k   =   Boltzmann's distribution coefficient

It can be seen that the intensity of the emitted radiation is proportional to the number of excited atoms. For a given total number of atoms, the number of excited atoms is a function of the temperature and the excitation energy E. As illustrated in Eq. (1.1), E equals $h\nu$; hence, the number of excited atoms can be related to the energy, and hence to the frequency of the photons emitted.

Equation (1.3) tells us that the intensity of emission increases when the temperature of the system increases. It also says that the intensity of emission decreases when the energy of excitation increases.

It can be understood that as the temperature increases, more atoms become excited and, therefore, more are available for emission. But as E—the energy necessary to cause excitation—increases, fewer atoms will become excited at any particular temperature and so the intensity of emission will decrease.

The energy of excitation can be found by measuring the frequency of the radiation emitted as indicated by Eq. (1.1).

Sometimes it is convenient to measure radiation in terms of wavelength rather than frequency. This can be done by using Eq. (1.2).

Table 1.1

The Relationship Between Temperature, Excitation Wavelength, and the Number of Atoms in a Given Atom Population

Excitation Wavelength, Å	Number of excited atoms per unit population at		Enhancement caused by 500°K temperature increase
	3000°K	3500°K	
2000	10^{-10}	$10^{-8.6}$	30
3000	$10^{-6.7}$	$10^{-5.7}$	10
6000	$10^{-3.3}$	$10^{-3.0}$	2.3

The relationship between emission intensity, temperature, and wavelength of the emitted light is demonstrated in Table 1.1. As the number of excited atoms increases, the intensity of radiation increases. It can be seen that there is a relationship between the number of excited atoms (and hence spectral intensity) and change of wavelength and change of temperature of the system. It can be readily understood, therefore, that the intensity of emission lines is sensitive to the wavelength range of the spectrum being considered and the temperature of the system. Increased temperature increases spectral intensity, increased wavelength increases intensity.

In contrast, the degree of absorption in atomic absorption is given by

$$\int_0^\infty K\nu \; d\nu = \frac{\pi e_2}{mc_2} \; Nf \tag{1.4}$$

where

K = absorption coefficient at frequency ν
e = charge of an electron
m = mass of an electron
N = number of absorbing atoms
c = speed of light
f = oscillator strength of the absorption line

The oscillator strength of a line is the probability of a transition between the ground state and the excited state involved. In emission spectrography the oscillator strength is related directly to the intensity of emission. If one emission line is twice as strong as another emission line from the same spectrum, then the oscillator strengths of these two lines are in the ratio 2:1. The greater the oscillator strength, the more likely is a transition to take place. The oscillator strength can be defined mathematically as

$$f = \frac{\lambda_0^2}{8\pi} \frac{g_2}{g_1} A_2^{1-} \tag{1.5}$$

where

λ_0 = wavelength of the resonance line
A_2^1 = Einstein's coefficient of spontaneous emission

$\dfrac{g_2}{g_1}$ = statistical weights of atoms in states g_1 and g_2

It can be readily seen that f, the oscillator strength, is a physical property of the transition of a particular element, and is therefore not a variable under normal conditions.

If we look back at Eq. (1.4), we see that the degree of absorption given by $\int_0^\infty K\nu \, d\nu$ is equal to a number of constants, that is, to e, m, c, and f multiplied by the total number of atoms in the light path.

It is very important to remember this relationship, because it demonstrates one of the fundamental differences between atomic emission and atomic absorption. It should be noted that temperature, t, and the energy of transition, E, are not part of the mathematical relationship governing the degree of absorption. Hence, to a first approximation the total degree of absorption is equal to a constant times the number of free atoms in the light path. This can be represented as Eq. (1.6):

$$\text{total absorption} = \text{constant} \times \text{Nf} \tag{1.6}$$

1.2 THE BEER-LAMBERT LAW

The absorption by atomic systems follows Beer's law, which is

$$I_1 = I_0 \, e^{-a,b,c} \tag{1.7}$$

where

I_1 = amount of light emerging from a solution
I_0 = intensity of light falling on a solution
a = absorption coefficient
b = path length
c = concentration of the solution

The Beer-Lambert law is one of the most fundamental laws relating the degree of absorption by a solution with the concentration of its components. The ramifications of this law will not be

explored in this book, but may be studied elsewhere if necessary
[1].

It may be of interest to note that the law can often be used
in its simple form,

$$A = a, b, c \qquad\qquad (1.8)$$

where A is the total amount of light absorbed and a, b, and c are
defined as in Eq. (1.7).

Although the basic principles of the Beer-Lambert law are
applied to atomic absorption spectroscopy, in practice it is not
feasible to use this relationship in the same way as it is used for
the measurement of solution concentrations. This is because solutions analyzed are invariably homogeneous, and molecular concentrations are therefore constant throughout the sample absorption
light path. However, in a system of free atoms, the concentration
of free atoms is not constant throughout the absorption light path.
Hence the Beer-Lambert law cannot be used directly to determine
the concentration of an atom generated from sample solution. Instead it is necessary to use Eq. (1.5) relating the degree of absorption with the total number of atoms in the light path.

As will be discussed later, the number of atoms in the light
path is in a state of dynamic equilibrium with the sample solution
and with the products of combustion. This number of free atoms
is in turn proportional to the concentration of the metal being
determined in the sample. We can therefore construct calibration
curves relating the degree of absorption and the concentration of a
solution. It can be seen from Table 1.1 that the number of excited
atoms varies considerably with a change in temperature of the
system. It can also be seen that the number of excited atoms represents a very small proportion of the total number of free atoms in
that population. It can be concluded from this that the great bulk
of free atoms exist in the ground state (or the unexcited state) at
temperatures normally encountered in flame atomizers. These unexcited atoms do not contribute to the emission signal, but they
can contribute to the degree of absorption because they are all
capable of absorbing energy. Table 1.1 also reveals that even if the
number of excited atoms increases greatly, the total number of
unexcited atoms varies almost imperceptibly. For a given popula-

tion of atoms the ground state is the most highly populated and within experimental error is independent of temperature. Since the degree of absorption is a function of the number of free atoms that can absorb energy, we can conclude that the degree of absorption is virtually independent of temperature for a given atom population. It can also be seen from Eq. (1.4) that there is no term relating the absorption wavelength to the total absorption. In contrast to emission, the total absorption is independent of the absorption wavelength.

1.3 ADVANTAGES OF ATOMIC ABSORPTION SPECTROSCOPY

To summarize, based on physical properties of free atoms, atomic absorption spectroscopy has the following inherent advantages over atomic emission.

1. Atomic absorption is virtually independent of the temperature of the system, if we ignore the effect of temperature on atomization efficiency, which is itself a variable affecting both emission and absorption.

2. Atomic absorption is independent of the wavelength of the absorption line.

3. The great majority of free atoms exist in the ground or unexcited state and contribute to the atomic absorption signal. In contrast, the atomic emission signal depends directly on the number of excited atoms in the system, which are only a small fraction of the total number of atoms.

However, it must be stated that the emission signal is measured as a small signal against a background that is virtually zero. It is possible to amplify this small signal many times and therefore compensate to some degree for the loss of absolute signal encountered in emission spectrography.

4. In atomic absorption, it is only necessary to measure I_1 and I_0 in order to measure absorption. It is frequently easier to measure this ratio rather than to measure absolute quantities, particularly when the effect of interferences must be corrected for in order to obtain accurate results. It should be noted at this point that, when flame atomizers are used, changes in flame temperature can affect the efficiency of producing atoms from a given sample. This directly controls N_1, the number of unexcited atoms in the

light path. Such a temperature effect on atomic absorption can be readily observed experimentally. However, it should not be confused with the effect of temperature on the physics of the absorption process by each atom, which is generally negligible.

When carbon filament or carbon tube atomizers are used, the efficiency of atomization is vastly improved over that encountered in flames. Because of this tremendous increase in efficiency, extremely high sensitivities—of the order of 10^{-13} g—have been achieved by atomic absorption spectroscopy using relatively simple equipment.

Although it has been claimed that such low concentrations can be measured by atomic emission, the difficulties involved in controlling the equipment so as to provide extremely low noise levels and very high amplification are formidable. This has hitherto precluded the use of flame emission as a routine instrument for analysis of concentrations at this low level. However, analyses at this level are becoming increasingly commonplace using atomic absorption spectroscopy.

1.4 COMPARISON OF ATOMIC ABSORPTION AND ATOMIC EMISSION

Difference in Basic Principle

The basis of atomic emission spectroscopy is that a small emission signal is measured over and above a background emission that is as small and steady as possible. The emission intensity is a measure of the concentration of the elements in the sample causing the emission. The sensitivity is generally limited by the noise level, or the short-term fluctuations in signal intensity of both the emission signal and the background signal.

In atomic absorption spectroscopy the sensitivity is defined as that concentration of sample which will give 1% absorption. In this instance, a strong radiation signal goes through the atomic population and radiation is absorbed. At the limits of sensitivity, a small change in radiation intensity is measured. The signal is generally the difference between an intense unabsorbed signal and a somewhat less intense signal after slight absorption. The definition of sensitivity does not take into account the noise level of a signal. In practice, this is not a misleading definition because generally it is not difficult to achieve the sensitivities derived using this definition.

One of the advantages of this type of definition is that it is general in its application. No improvement in amplification or smoothing of the signal to give lower noise level will change the sensitivity of the analytical method. Hence, research workers and routine operators can communicate data to one another with more certainty that their data can be achieved by other workers. Hence, the sensitivity data given in the literature is generally obtainable in the routine laboratory. In contrast, in atomic emission, sensitivity data are sometimes quoted from research labs using very sophisticated equipment with high amplification and low noise levels. This type of equipment is capable of achieving an analytical sensitivity that is not always achievable on normal commercial equipment.

When flame atomizers are used, similar sensitivities are achieved using either atomic emission or atomic absorption, providing that the spectral line used has a long wavelength. An example of such an element is sodium 5890 Å. On the other hand, if the spectral line is of short wavelength, such as zinc 2138 Å, very strong absorption signals are obtained but the emission signals are extremely weak. For these elements, atomic absorption is much more sensitive as an analytical technique than atomic emission.

In recent years carbon atomizers have been used that are capable of providing sensitivity down to 10^{-12} or 10^{-13} g. These atomizers are more sensitive than atomic emission for virtually all of the metallic elements that can be analyzed by atomic absorption. This technique will be discussed in Chapter 7.

Interference

Interference in atomic emission arises from several sources.

Molecular Interference. Any broad-band emission at the same wavelength as the atomic emission line contributes to the background. Unless special precautions are taken to remove this interference or correct for it, it will result in a direct error in the atomic emission signal. Typical examples are molecular emission arising from hydroxyl ions and other molecular sources

Atomic (Spectral) Interference.   Any element that emits at the same wavelength as the line being measured for the sample element is a direct source of error. In atomic emission many spectral lines are observable for most elements, particularly the transition elements. Such overlapping lines can be a direct source of error if they coincide with the wavelength of the spectral line being measured.

Chemical Interference.   The chemical form of a sample directly affects the efficiency of atomization and excitation. Stable compounds are difficult to decompose, and therefore atomization efficiency is decreased, resulting in a decrease in atom population. This results in a loss of emission signal for a particular concentration of the element being analyzed. A correction must be made in order to achieve accuracy.

In atomic absorption spectroscopy both molecular and atomic spectral interference can be eliminated by using modulated equipment (see Section 2.7). However, chemical interference is caused by the total atom population and therefore affects atomic absorption as much as atomic emission. If chemical interference is suspected, it is frequently necessary to extract the element being analyzed into a matrix that eliminates the interference (the same matrix must be used in the sample and in the standard). This approach can be taken for both emission and absorption.

Limitations on Use

One of the difficulties encountered in atomic absorption spectroscopy is that generally it is possible to determine only one element at a time. This is because it is necessary to use a radiation source that emits the correct wavelength for absorption, for which a hollow cathode is used. Commercial instruments are available that analyze two or three elements at a time, but this is generally the limit for routine work. In contrast, the emission method can be used for many elements simultaneously.

Both atomic absorption and atomic emission are limited in general to the metallic elements of the periodic table. The resonance lines for nonmetallic elements are generally in the vacuum ultraviolet and are not detectable on normal commercial equipment. Some upper excitation lines are available for atomic emission, but generally these are not reliable for routine analytical purposes.

1.5 CONCLUSION

In conclusion, it can be said that atomic absorption spectroscopy provides an analytical technique at least as sensitive as other elemental analytical techniques, and recent innovations make it much more sensitive for quantitative analysis. In general, it is not subject to as many errors as emission methods, although it is not entirely free from interferences.

Its high degree of freedom from interference effects makes atomic absorption spectroscopy a usually reliable, accurate method of analysis capable of high precision without requiring highly trained personnel or very specialized equipment.

Its use for qualitative analysis is limited, but the method provides an excellent quantitative technique.

REFERENCE

1.   J. W. Robinson, *Undergraduate Instrumental Analysis*, 2d Ed., Marcel Dekker, New York, 1973.

Chapter 2

EQUIPMENT

2.1 THE OPTICAL SYSTEM

Atomic absorption spectroscopy uses the same basic optical system
as other forms of absorption spectroscopy. The components include a radiation source, a monochromator to select the desirable
wavelength of radiation to be examined, a sample container, and a
detector for measuring the intensity of radiation after it passes
through the sample. The components can be lined up in two basic
optical systems: (1) single-beam optics and (2) double-beam optics.

The Single-Beam Optical System

A single-beam optical system is shown in Fig. 2.1. Radiation
from the light source passes through the simple atoms generated in
the atomizer. From here the light proceeds to a monochromator.
The monochromator disperses the radiation in the same manner as
a prism. By using a slit system with the monochromator, only light
of the desired wavelength is allowed to proceed along the light
path. This light passes to the detector. The detector in turn measures the intensity of the light, and the signal is read out on a
recorder.

A typical absorption signal obtained from a zinc sample is

15

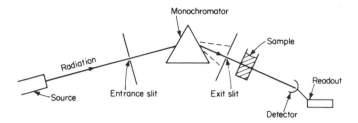

Figure 2.1. Schematic diagram of single-beam optical system. From Robinson [1].

shown in Fig. 2.2. In a typical experiment it may be seen that with no zinc present the light intensity is approximately 100 units. When zinc is introduced into the system, the signal intensity drops to approximately 95 units, showing a loss of 5 units of intensity. This is 5% absorption; that is, 5% of the light that entered the system was absorbed by the atoms.

The single-beam system works well in practice and is commonly used. However, it is subject to one major source of error. If the intensity of radiation from the source varies, then the entire signal drifts, as illustrated in Fig. 2.3.

In this case it can be seen that there is a significant change in the intensity of radiation reaching the detector. If the entire signal is being recorded, then it is not difficult to observe and correct for this drift; but if the signal intensity is read out on a digital readout system or a simple needle, then an error results. As another example, if the background drops to an intensity of 93, and there

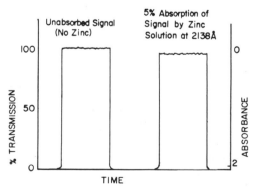

Figure 2.2. Absorption of 5% of signal by zinc solution.

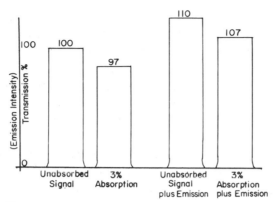

Figure 2.3. With background emission of 10 units, the signal after absorption appears to be 107 units, which is greater than the real unabsorbed signal (100).

is a 5% absorption by the sample, the reading would be approximately 88. An unsuspecting operator could mistake this for 12% absorption, since he would be under the impression that the original intensity was 100.

This source of error can be largely overcome by using the double-beam system.

The Double-Beam Optical System

A schematic diagram of the double-beam system is shown in Fig. 2.4. In this system the light from the radiation source is split

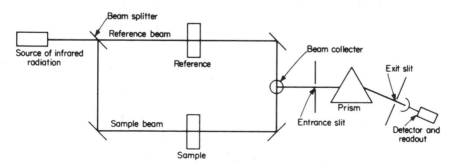

Figure 2.4. Schematic diagram of a double-beam optical system. From Robinson [1].

by a beam splitter and forms two paths, the reference path and the
sample path. The beam splitting alternately directs the light from
the radiation source along the reference path and then along the
sample path. The light beams passing down these two paths are
equal in intensity and pass alternately through sample and a reference. After passing through the reference and sample they are then
recombined and continue through a monochromator split system
to the detector.

In practice, the reference beam either is not absorbed (e.g., by
using an empty reference beam) or is absorbed by a constant quantity by a standard reference cell. It is not possible to use a reference
flame. The radiation in the sample beam, however, is absorbed by
the atomized sample. When the two beams are recombined, an
oscillating signal is produced which falls on the detector.

If there is no absorption by the sample or by the reference,
then the two beams recombine to form the original unsplit beam.
This beam does not vary in intensity as the detector views alternately the reference beam and the sample beam. In this instance,
the readout gives zero absorption.

When there is absorption in the sample beam, then the signal
reaching the detector varies accordingly. Typical examples are
shown in Fig. 2.5.

It can be seen that the amplitude of the signal falling on the
detector is greatly increased as the degree of absorption is increased.

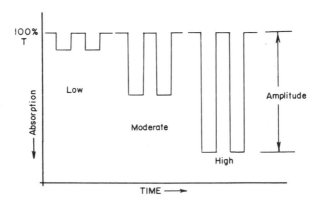

Figure 2.5. Typical signal reaching the detector with low, medium, and
high absorption.

In this instance, the amplitude is the difference in signal intensity between the reference and the sample beam.

The amplitude does not change appreciably when the intensity radiation from the source changes. For example, if the source intensity decreases 5%, then both reference and sample beam decrease 10% and there is no change in signal; that is, both beams still read almost the same.

The great advantage of this system is that drift or small changes in intensity of the radiation source do not cause a signal error in the readout. This can be seen in Fig. 2.6, which illustrates a signal with no drift in the radiation source and a 5% drift in the radiation source.

Note that there is very little change in the amplitude of the signal falling on the detector.

One of the difficulties with atomic absorption spectroscopy is that a suitable reference is not available for use in a double-beam system. All atomizers are dynamic in operation, and it is very difficult in practice to make an atomizer that will give a faithful reproduction of the background signal found in the sample atomizer. In practice, some commercial equipment still uses the single-beam system and relies on stabilized radiation sources and repeated checks in order to eliminate errors caused by drift. Other systems use a double-beam system that includes a reference beam, but not as a reference cell. In other words, the reference beam is empty, and in this case we are really using a "pseudo" double-beam system.

Each optical system has worked well in operation and has been used to obtain reliable data. The single-beam system requires

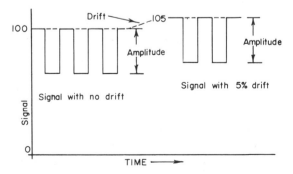

Figure 2.6. Change of signal with baseline drift.

controlling the intensity of radiation from the hollow cathode over extended periods of time. This can best be done by using a highly controllable power supply to the hollow cathode system, which is frequently quite expensive. In contrast, when the double-beam system is used, some drift in the hollow cathode intensity and hence in the power supply to the hollow cathode can be tolerated without loss of analytical sensitivity. The double-beam system, therefore, enables one to use less expensive equipment for the hollow cathode and power supply unit. However, one must also use the extra optical equipment necessary to incorporate a double-beam system. We are, therefore, faced with a common dilemma in spectroscopy: whichever way you go, some advantages are gained and some advantages are lost. The choice is left up to the operator.

      We shall now turn to the individual components of the equipment.

2.2 ABSORPTION LINE WIDTH

The natural spectral width of atomic absorption lines is about 10^{-4} Å. This finite width reflects the natural energy spread of the ground state and the excited state under ideal conditions in the atoms of interest. The natural line width is broadened by the Doppler effect, which causes an increase in the absorption line width with increasing temperature of the atoms. It is common for absorption lines to be increased to about 0.01 Å by this effect. Collision between atoms causes a broadening of the energy levels in the ground state and the excited state and, hence, a broadening of the absorption bands. This is called pressure broadening and can be caused by collisions between like atoms or different atoms (the latter is called Lorenz broadening). It is common for pressure broadening to increase the line width to 0.02 Å. In addition, absorption line widths are dependent on the wavelength of the resonance line or absorption line. If the resonance line is at a long wavelength, the absorption line widths are greater than if they are at short wavelengths. This is illustrated in Table 2.1, where are shown the absorption line widths for sodium and potassium, representing short and long wavelengths and temperatures of 1000 and 3000° K.

Table 2.1

Spectral Widths of Absorption Lines[a]

| | | Absorption line width | |
		1000°K	3000°K
Na	5895 Å	0.028 Å	0.048 Å
Zn	2138 Å	0.006 Å	0.01 Å

[a]Fundamental line width broadened by temperature and pressure.

If a continuous source such as a hydrogen lamp or a tungsten lamp were used for atomic absorption, only the energy in the narrow absorption wave band indicated above would be absorbed. With a normal commercial instrument, a spectral wave band of about 1 Å falls on the detector. Normally, atoms absorb over a range of about 0.02 Å. If all of the light in this absorption bandwidth were absorbed from an emission wave band of 1 Å, the detector would record the loss of only 2% of the signal falling on it. The rest of the signal would not be absorbed by the free atoms because it is not included in the absorption line width. The sensitivity of an atomic absorption method is defined at that concentration which will give 1% absorption of the resonance line (width 0.02 Å). In this instance, using a continuous light source, the total energy falling on the detector would diminish only 0.02% and the detector would register a decrease in signal of the same magnitude. Conventional detectors are not capable of measuring such small changes in signal with any reasonable degree of precision. Therefore, it is possible to achieve either high sensitivity or useful quantitative data if a continuous source is used, particularly at shorter wavelengths.

This difficulty was overcome by Walsh [2], who demonstrated that hollow cathodes emit very narrow spectral lines. If the hollow cathode is made of the same element as the sample being analyzed, the line width of the hollow cathode radiation is somewhat narrower than the atomic absorption line width and, therefore, completely available for absorption. In addition, since there is usually virtually no background light falling on the detector, the signal approaches zero energy when the spectral line width is completely absorbed.

In summary, it can be said that atomic absorption line widths are very narrow and cannot be easily measured with conventional equipment using a continuous light source. A hollow cathode is necessary for operation.

2.3 THE HOLLOW CATHODE

A schematic diagram of a hollow cathode, based on the design of Jones and Walsh [3], is shown in Fig. 2.7. In practice, the voltage is applied between the anode and the cathode, the anode being negative and the cathode positive. The system is filled with a noble gas, such as helium, argon, or neon. The filler gas, e.g., argon, is ionized by the anode and becomes a negative argon ion. The negative ion is attracted by the positive cathode and accelerated under the influence of its charge. When it reaches the cathode it impinges on the metal surface, dislodging or "sputtering" excited metal atoms into the space inside the cathode. The excited metal atoms emit radiation of characteristic wavelengths and return to the ground state. The emitted radiation is used as the light source for the atomic absorption system. After the atoms return to the ground state, they form a cloud of free atoms and diffuse slowly either to the walls of the cathode or to the glass walls forming the envelope. If the cathode is hot, spectral line broadening occurs caused by the Doppler effect. This is detrimental to the successful operation of the instrument, since the "wings" of these emission lines cannot be absorbed by the free atoms from the sample.

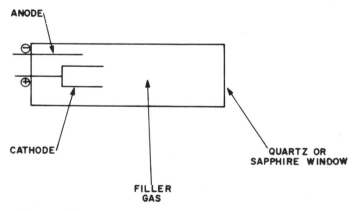

Figure 2.7. Schematic diagram of a sealed hollow cathode.

Another problem with the cold atom cloud inside the hollow cathode is that it is able to reabsorb some of the radiation emitted by the cathode itself. This absorption is at the very center of the emission line and leads to distorted calibrated curves because not all of the lines emitted by the hollow cathode remain to be abosrbed by the free atoms of the sample. As will be described later in this section, this problem can be somewhat alleviated by using a demountable hollow cathode whereby the atom cloud is continuously removed by pumping.

An illustration of the effect of absorption by the atom cloud on the signal emitted from a hollow cathode is shown in Fig. 2.8.

Intensity of Radiation Emitted from Hollow Cathodes

For a single-element hollow cathode, the intensity of the emitted radiation depends on the effectiveness of the bombardment of the hollow cathode by the charged filler gas to cause sputtering. The kinetic energy of an ion of bombarding filler gas must be greater than the energy necessary to dislodge the metal from the

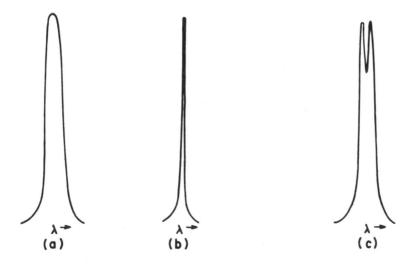

Figure 2.8. Distortion of spectral line shape in a hollow cathode. (a) Shape of spectral line emitted by a hollow cathode. (b) Shape of spectral energy bond absorbed by cool atoms in a hollow cathode. (c) Shape of net signal emerging from a hollow cathode.

cathode surface, that is, the lattice energy. The lattice energy of the metal is a physical property of the metal and cannot easily be varied except by temperature. But, as we have already seen, it is better to operate the hollow cathodes at low temperatures to avoid Doppler broadening. Hence, we are limited to low-temperature operations and we cannot take advantage of the variable.

The kinetic energy of the bombarding ion is directly controlled by the mass of the ion, the voltage across the electrodes, and the number of collisions per unit time that the ion experiences on its way to the cathode. The mean free path of the ion determines the duration of acceleration, and therefore its mean kinetic energy.

In practice, it is important that the geometry of the system, the mean free path of the ion (and therefore its pressure), and the voltage across the anode and cathode be carefully controlled in order to obtain a steady signal. The optimum voltages and pressures have been determined. It has also been found that the best shape for the hollow cathode is a tube sealed at one end and open at the other. This tube shape tends to keep free atoms inside the cathode, and therefore extends the lifetime of the hollow cathode system. In addition, it provides an extended area over which atoms are emitted, and therefore increases the radiation intensity of the hollow cathode system.

Filler Gas

The most common filler gas used is argon, because it is the cheapest and because it has a fairly simple spectrum. In addition, argon does not easily "clean up" inside the hollow cathode, and therefore provides a system with a reasonable long life. Since argon emits at characteristic wavelengths, there are occasions when an argon emission line coincidences with, or is very close to, the resonance line of the element being examined. This is so with lead hollow cathodes, for which helium is recommended rather than argon in order to avoid overlap of emission lines. Neon has been used in some cases, but the added expense of using this gas usually precludes its use. Leakage of air into the system will allow oxygen or nitrogen to become ionized rather than the filler gas. These ions are reactive, and they result in a decrease in sputtering. This prevents operation of the system and must be avoided.

Loss of Metal

The metal used in the hollow cathode is the source of radiation of the resonance line. With time this metal will all "sputter" away, leaving only the support material, which does not emit at the desired wavelength. The sputtering can be slowed by operating at low voltage. This is particularly important when volatile metals such as lead, tin, or arsenic are used. Commercial manufacturers have spent significant research time on this problem, and have come to recommend conditions whereby a suitable emission intensity can be obtained while prolonging the lifetime of the hollow cathode system. It can be readily appreciated that increasing the voltage of the hollow cathode will increase its intensity but will also decrease its lifetime.

Hardening

The free atoms from the hollow cathode diffuse out into the open system and slowly trap the filler gas between themselves and the glass envelope. When this occurs the pressure inside the cathode decreases until it becomes too low for the system to work. Under these conditions it is said that the hollow cathode has "hardened." If the trapped filler gas can be released or if more gas can be introduced into the hollow cathode system, then the hollow cathode can be rejuvenated. With sealed systems, however, this is seldom possible.

Other Hollow Cathode Types

High-Intensity Lamps. Improvements on the original hollow cathode design have been made to provide high-intensity lamps. This provides a calibration technique with less self-reversal, and hence with better analytical results. The first high-intensity hollow cathode was developed by Jones and Walsh [3] and is illustrated in Fig. 2.9. The anode and cathode are similar in shape and design to those used in the original instrument, but two additional cathodes are placed near the mouth of the hollow cathode. A current is passed between these auxiliary electrodes. The electrons evolved

cause increased excitation of the atom cloud at the point of intersection, greatly increasing the intensity of the resonance lines of
the element. It is also found that unwanted ionization lines are
suppressed.

The high-intensity lamp allows the detector to operate at
lower voltage, and therefore to provide a quieter signal and per-
mitting better analytical data to be achieved.

Two other designs have been provided by Westinghouse, which
used (a) the open hollow cathode and (b) the shielded hollow
cathode. These are illustrated in Fig. 2.10. The open hollow cathode
is similar to Walsh's original design. The shielded hollow cathode
forces the carrier gas ions to the inside of the cathode, causing

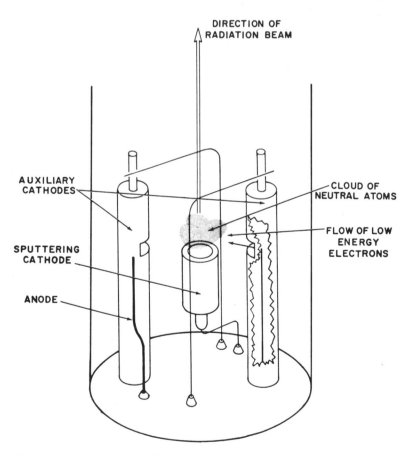

Figure   2.9.   Schematic diagram of a high-intensity hollow cathode.

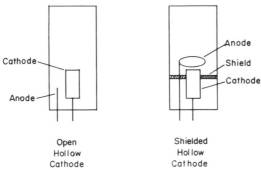

Figure   2.10.   Sealed hollow cathode designs.

enhancement of the resonance line. Typical emission lines of the open and shielded hollow cathodes and of Walsh's high-intensity lamp are shown in Fig. 2.11. It can be seen that the intensity of the nickel resonance line is increased while the intensity of the other nearby lines is decreased. Typical calibration curves obtained from these hollow cathodes are shown in Fig. 2.12. With the open and shielded hollow cathodes, nearby lines cause the calibration curves to deviate because of the presence of unabsorbable radiation. However, the problem is greatly decreased with a high-intensity lamp, as can be seen in Fig. 2.12.

   The Barnes Demountable Hollow Cathode.   One of the difficulties with the sealed hollow cathode lamp is that generally it is useful only for work with the element from which the cathode itself is made. Hence, if it is necessary to determine ten different elements, then ten different hollow cathodes are necessary. This is

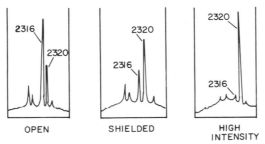

Figure   2.11.   Typical emission lines of open, shielded, and high-intensity hollow cathodes. Note the increasing intensity of the absorbing 2320-Å line and the decreasing intensity of the nonabsorbing 2316-Å line.

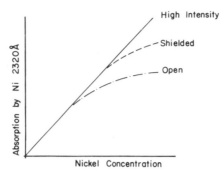

Figure 2.12. Calibration curves for nickel vs. absorption, using different hollow cathodes.

an expensive and time-consuming project, particularly in a research-oriented lab, where a number of different elements must be analyzed in the course of a day. The problem can be greatly alleviated by using the Barnes demountable hollow cathode lamp, which is illustrated in Fig. 2.13. The advantage of the system is that the cathode itself can be easily removed and replaced by a cathode of another element in a few minutes. The system can be pumped down and operating again within 15-20 min. The filler gas flows constantly under controlled pressure through the system. Atom clouds are not allowed to build up, and hence self-absorption is greatly decreased. The initial outlay to buy the demountable hollow cathode is comparatively high, but since it is so easy and inexpensive to replace a cathode, the cost decreases over an extended time period. In addition, it is very convenient to be able to change cathodes or filler gas with little difficulty.

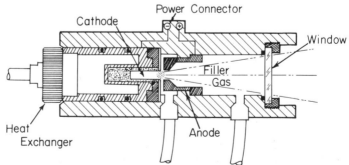

Figure 2.13. Barnes demountable hollow cathode lamp.

Multi-Element Hollow Cathodes. Frequently, it is desirable
to analyze two or three elements at the same time. This is a difficult procedure using single-element hollow cathodes. The problem
can be overcome by using a multi-element hollow cathode. The
first such hollow cathode was made from iron, nickel, and vanadium.
It is illustrated in Fig. 2.14. In practice, it was found that emission
spectra from all three elements were indeed generated and the
resonance lines used for simultaneous analysis. However, it was
found that inevitably one element was more volatile than the other
two elements. With time, this element tended to cover over the
surface originally occupied by the other two elements. Consequently the intensity of the radiation from this element increased, and
the intensity of radiation from the other two elements decreased.
The lifetime of the multi-element hollow cathode was thereby
greatly reduced. In addition, the steady drift in intensity presented
analytical problems. At least some compatibility in the volatility of
the metals is necessary or the lifetime will be too brief to be useful.
    Subsequent to the original design, shown in Fig. 2.14, the use
of alloys and powder metallurgy has increased greatly the choice of
metals used in multi-element hollow cathodes. Although their lifetime is short compared to that of a single-element hollow cathode,
at times they are very valuable for particular analytical requirements.

2.4 THE MONOCHROMATOR AND SLIT SYSTEM

The function of the monochromator slit system is to isolate radiation of the desired wavelength from the rest of the radiation emitted
by the light source. The desired radiation, which is usually the

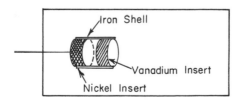

Figure 2.14. Multi-element hollow cathode built from iron, nickel, and
vanadium.

emitted resonance line of the element being analyzed, is permitted to travel down the light path to the detector, while all other radiation is prevented by this system from reaching the detector.

The two major components involved in this isolation are the slits and the monochromator. These are illustrated in Fig. 2.15. It can be seen from Fig. 2.15 that there are two slits, an entrance slit and an exit slit. The entrance slit permits radiation from the hollow cathode to reach the monochromator, but prevents all other radiation such as stray radiation from lights, windows, etc., from continuing down the light path. The exit slit, on the other hand, is placed after the monochromator, and its function is to permit only radiation of the correct wavelength to pass, all other radiation being prevented from passing.

In practice, the slits are usually made of two parallel knife edges between which radiation can pass. Often the distance between the edges of the slits is variable, but on some systems this slit width is fixed. In the exit slit the distance between the slits controls the wavelength range that is permitted to pass through. The closer the slits are together, the narrower the permitted wavelength range will be. This wavelength is called the spectral slit width. In atomic absorption spectroscopy the resonance line emitted by the source is very narrow and there are frequently no other lines in the vicinity of this emitted line. Consequently, in many cases low-resolution equipment is all that is necessary. There are cases, however, where high resolution is definitely an advantage, particularly with the transition elements where the emission spectra

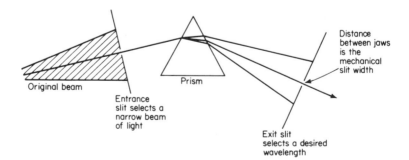

Figure 2.15. Schematic diagram of a monochromator slit system. From Robinson [1].

are very rich. Commercial equipment is designed to operate sufficiently well to meet all resolution requirements.

The function of the monochromator is to disperse the radiation that falls upon it according to wavelength. It operates in the same manner as a prism that separates the sun's light into a rainbow.

It will be remembered that the hollow cathode emits a number of very narrow lines and that the background between these lines is virtually zero. It is necessary to separate the resonance line from other spectral lines in its immediate vicinity. These lines originate from the metals in the hollow cathode and the filler gas. The spectra have long been documented, and the wavelengths of these lines are quite predictable.

The analytical sensitivity of atomic absorption is defined as that concentration of metal which will diminish the signal by 1%. If the intensity of the signal is I_0 before absorption and I_1 after absorption, the sensitivity limit can be defined as

$$\frac{I_0 - I_1}{I_0} = \frac{1}{100} \tag{2.1}$$

If, however, a second line of intensity I_0' falls on the detector and is not absorbed, the same concentration that originally gave 1% absorption will now give absorption equal to

$$\frac{(I_0 - I_1) + I_0'}{I_0 + I_0'} \tag{2.2}$$

The loss of sensitivity depends on the relative intensity of I_0 and I_0'. In order to achieve maximum sensitivity it is better to eliminate all sources of I_0'.

The principal types of monochromators used in atomic absorption spectroscopy are the prism, the grating, and the filter.

Prisms

It is not generally possible to use glass prisms in atomic absorption spectroscopy because glass is not transparent at wavelengths less thån 3500 Å. In practice, therefore, quartz prisms are preferred.

The resolution required to separate two lines is defined as

$$R = \frac{Av\ \lambda}{\Delta\lambda} \qquad\qquad (2.3)$$

where Av λ is the average wavelengths of the two lines and $\Delta\lambda$ is the difference in wavelength between the two lines. In other words, to separate two lines of wavelengths 5000 and 5002, the required resolution would be 5001/2 = 2500.5.

The resolution of a prism is given by the equation

$$R = \frac{\delta\eta}{\delta\lambda}\cdot t \qquad\qquad (2.4)$$

Where $\delta\eta$ is the rate of change of refractive index with change in λ and t equals the thickness of the base of the prism.

The rate of change of refractive index of the wavelength depends upon the wavelength itself and varies throughout the spectral region. However, most prisms used in atomic absorption are adequate for the work, since only low resolution is usually required.

Gratings

A grating consists of a flat metal surface onto which parallel grooves have been cut, all equidistant from each other. The grooved surface reflects and resolves the light falling upon it. Reinforcement of the light occurs when

$$\delta\eta = d(\sin i \pm \sin \theta) \qquad\qquad (2.5)$$

where

η = a whole number (the order)
λ = wavelength
d = distance between grooves
i = angle of incidence of light on the grating surface
θ = angle of dispersion of radiation wavelength λ

When d is decreased, then the useful wavelength range λ of the grating is decreased. Hence when the number of lines on the grating is increased, the useful wavelength range is decreased.

In practice, original gratings are expensive and seldom used. A replica grating, made in much the same way that records are made, is quite satisfactory and far less expensive.

The grating is the most popular type of monochromator used in atomic absorption spectroscopy for several reasons, the most important of which is that the grating is durable, its dispersion is constant over its useful spectra range, and suitable gratings are now reasonable in cost.

The resolution of a grating is given by

$$R = N \times n \tag{2.6}$$

where N is the total number of lines rolled on the grating and n is the order. It can be seen that resolution increases with the number of lines rolled on the grating. Suitable commercial gratings are readily available for this work.

Filters

Light filters that pass only a limited spectral range have been used in atomic absorption spectroscopy. One difficulty with this type of filter is that the spectral range it permits to pass is quite wide. The most successful filters are interference filters. Part of the light beam passes directly through and part is reflected by the surfaces of the filters. If the thickness of the filter is d, then $\eta\lambda = 2d$ reinforcement takes place and the light is permitted to pass. Light of other wavelengths is lost by interference effects. The system is shown in Fig. 2.16. These are available only for the longer wavelength ranges used in atomic absorption spectroscopy.

2.5 DETECTORS

By far the most common detector used in atomic absorption spectroscopy is the photomultiplier. Figure 2-17 shows a schematic

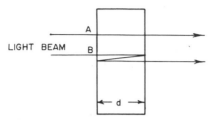

Figure 2.16. Interference filter: Reinforcement occurs when the extra distance traveled (2d) by beam B is a whole number of wavelengths (nλ).

diagram of a photomultiplier. It can be seen that the photomultiplier is a series of electrodes, each with a potential positive to the previous electrode. When a photon hits the first emissive surface, an electron is ejected and attracted to the next dynode. It is accelerated in the process, and when it hits the next dynode it ejects several electrons. These in turn are attracted to the following dynode, and they in turn each eject several other electrons. This process is continued through each dynode, and a shower of electrons arrives at the final gathering post. In this manner the single photon is responsible for generating a significant electrical signal. The sensitivity of this system, which is in fact a simple amplification system, is dependent on the voltage between the dynodes. The greater the voltage difference, the greater the amplification. However, if the voltage is increased too much, the signal becomes erratic and the output becomes "noisy." Typical noisy and acceptable signals are shown in Fig. 2.18. In practice, the photomultiplier is operated at the maximum voltage where it is not noisy. This can be determined manually

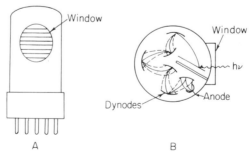

Figure 2.17. Schematic diagram of a photomultiplier. A. Side view. B. Top view showing configuration of dynodes and collector.

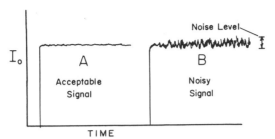

Figure 2.18. Emission signal intensity I_0 with (A) acceptable noise level, and (B) noisy signal.

by turning up the voltage until the signal becomes erratic, and then turning down the voltage until it is just quiet again.

The type of photomultiplier used depends on the wavelength of the radiation being monitored. Different photomultipliers use different photo-emissive surfaces to permit the energy of the photon to liberate electrons. A change in the surface changes the effective wavelength range over which the photomultiplier responds. The relationship between signal and wavelength for a radiation source of equal intensity is illustrated in a response curve. Typical response curves for the most common photomultipliers are shown in Fig. 2.19.

It can be seen that the output signal varies considerably with wavelength and hence with the photomultiplier used. It is essential that the correct photomultiplier be used, because if the wavelength range of the resonance line is outside the range of the photomultiplier, then erratic and unreliable analytical data will be obtained.

As indicated before, it is also necessary that the photomultiplier be operated at the correct voltage to provide a quiet signal.

2.6 ATOMIZERS

The function of the atomizer is to take the sample and reduce the metal being determined to the neutral atomic state. This is a very difficult procedure to reproduce and is at the very heart of the phenomenon of atomic absorption spectroscopy. All the major components of a typical commercial instrument are available at a quality level capable of providing maximum sensitivity and repro-

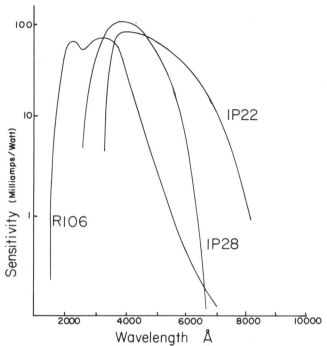

Figure 2.19. Response curves for some typical photomultipliers.

ducibility. However, this is not true with the atomizer. In general, atomizers are very inefficient and must be treated with great care in order to generate reproducible data. High sensitivity for qualitative detection is readily available with most commercial atomizers. The problems involved in qualitative work are spectroscopic in nature. However, the problem of reproducible quantitative analysis is a more difficult task and involves analytical chemistry. The most common atomizers in routine use are flame atomizers.

Flame Atomizers

The process of atomization in a flame will be discussed in Section 3.3. However, all atomizers have the same general basic function. A flame is formed using air, oxygen, or nitrous oxide as the oxidant and hydrogen, acetylene, coal gas, etc., as the fuel. A liquid sample is nebulized to liquid droplets, which are introduced into the base of the flame. Here the sample is decomposed and free atoms are generated. The free atoms find themselves in the hostile environment of the flame and are rapidly oxidized either to the metal oxide or to some other more stable molecular form. Flames are very inefficient atomizers. It can be calculated, for instance, that approximately one atom in a million in a sample is actually reduced to the free atomic state. All other atoms of the element being determined are not reduced to free atoms quickly enough and pass through the light path. They do not contribute to the atomic absorption signal. In recent years this problem has been overcome by using carbon atomizers with greatly increased sensitivity. The design and use of the carbon atomizer is discussed in Chapter 7.

There are two principal types of burners used for atomization. They are the total-consumption burner and the Lundegardh burner.

The Total-Consumption Burner. A schematic diagram of the total-consumption burner is shown in Fig. 2.20. The flame is created by the reaction between the oxidant, e.g., air, oxygen, or nitrous oxide, and the fuel, e.g., hydrogen, acetylene, town gas,

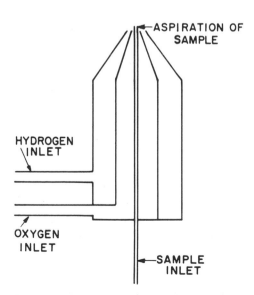

Figure   2.20.   Schematic diagram of a total-consumption burner.

butane, etc. Other exotic flames such as hydrogen-fluorine or oxy-cyanogen have been used, but not in routine analysis.

The sample is introduced into the base of the flame by aspiration created by the combustion process. The entire sample is introduced to the flame even if it includes suspended particles or solvent mixtures—hence the term total-consumption burner. The advantages of the total-consumption burner are as follows. (a) It is comparably simple to manufacture and, therefore, relatively inexpensive. (b) No fractionation of the sample can take place during aspiration. This eliminates sources of error by loss of nonvolatile samples. Its chief disadvantages include the following: (a) The analytical signal changes as the rate of aspiration of the sample into the flame changes. The aspiration rate is a function of the viscosity of the sample. The viscosity is changed by temperature changes or a change from a viscous solvent to a less viscous solvent. (b) Very viscous samples cannot be aspirated at all and so cannot be analyzed. They may be diluted with a less viscous solvent. (c) After introduction into the flame, the sample is broken into droplets. The droplet size depends on the chemical composition of the sample; that is, any factor that affects surface tension or viscosity will affect the sample droplet size. This in turn affects the atomic absorption signal obtained and therefore the analytical data achieved. Changes in surface tension can cause direct error in analytical data. (d) The burner tip can become encrusted with salts left after evaporation of the solvent. This can cause a change in aspiration rate of the sample and hence a direct error in the analytical results. Special precautions must be taken by the operator to be sure that the burner does not become encrusted during operation. (e) The burners are very noisy, both physically and electronically. The physical noise is very disturbing to the operator and is even more disturbing to other people in the lab. (f) The electronic noise contributes to poor reproducibility in the analytical data.

The total-consumption burner is used extensively in flame photometry. As with flame photometry, with good care, good results can be achieved in atomic absorption spectroscopy.

The Lundegardh Burner. The Lundegardh burner is quite different in shape and operation from the total-consumption burner. The basic design is illustrated in Fig. 2.21. The design incorporates two distinct consecutive steps, that is, nebulization and atomization of the sample. The burner uses the same oxidants and fuels as those used in the total-consumption burner. The oxidants and fuels are premixed in the barrel of the burner. The sample is aspirated into the same barrel, where nebulization and evaporation of the solvent takes place. The evaporated sample and small droplets are swept together with the combustion mixture into the base of the flame. Any components of the sample that are not nebulized or form large droplets collect on the side of the mixing chamber and are drained away.

The combustion mixture and vaporized sample enter the base of the flame. As can be seen from the schematic in Fig. 2.21, the flame is elongated, being approximately 10 cm long, and is formed in a shape of a slot. The light path is along the length of the burner.

The practical advantages of this burner are as follows. (a) The elongated flame introduces more atoms into the light path, increasing analytical sensitivity. (b) Burner encrustation is reduced because large sample droplets are eliminated from the system in the mixing chamber. (c) The burner is quiet to operate and therefore reduces irritation to the operator. (d) The analytical signal is significantly less noisy than the total-consumption burner because sample droplets are effectively removed from the light path and light scattering is greatly reduced.

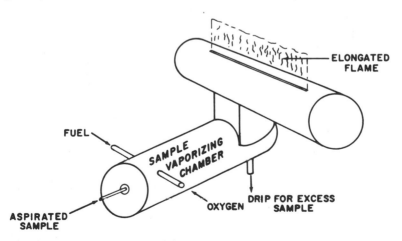

Figure 2.21. A nebulizer burner based on the Lundegardh burner (after the Perkin-Elmer 303).

However, the system has some limitations: (a) If the sample contains more than one solvent, any volatile material present is preferentially evaporated in the mixing chamber and the less volatile component drains off and does not reach the burner. It is possible under these circumstances to lose a significant portion of the sample element down the drain rather than sweeping it into the burner, resulting in loss of sensitivity and analytical inaccuracy. (b) The relatively large volume of the mixing chamber provides a hazard when volatile combustible samples are being analyzed. Special precautions must be taken if oxyacetylene flames are to be used because of the danger of flashback.

The nitrous oxide-acetylene flame has been used extensively for some applications of atomic absorption spectroscopy. It has been found in practice that the Lundegardh-type burner can be used for this flame quite successfully. Care must be taken to safeguard against flashback of the fuel-oxidant mixture. The manufacturers have done quite a good job in providing suitable safety precautions when the burner is used under these circumstances.

The Bowling Burner is another modification of the Lundegardh burner. This burner includes three slots instead of one as the orifice for supporting the flame. With this design the entire light beam in the optical system can be enclosed inside the flame, and an increase in sensitivity and accuracy results.

Beckman Corporation has introduced into the market a Lundehardh-type burner in which the nebulizing chamber is heated to 700-800°C. The increased temperature increases nebulization significantly, and an overall increase in analytical sensitivity results.

2.7 MODULATION OF EQUIPMENT

Atomic absorption is measured by the absorption of atomic resonance lines of the element being determined. The resonance lines are associated with transitions between the ground state and one of the low excited states of the atom in question. In flame photometry the atoms are excited first to the upper energy levels; then they fall to the ground state and in the process emit radiation. The transitions involved are the same transitions as those involved in absorption, so atoms emit at the resonance line—that is, at exactly the same wavelength at which they absorb. This presents a problem in measuring the absorption by the ground-state atoms. A schematic diagram of the equipment is shown in Fig. 2.22. It can be seen that the intensity of the source starts off at I_0, and after absorption the intensity drops to I_1. The absorption signal, there-fore, is $I_0 - I_1$. When the radiant energy emitted from the flame atomizer is added to this signal, however, the final radiation inten-sity falling on the detector after absorption is not I_1, but $I_1 + S$, where S is the intensity of radiation emitted at the resonance wave-length.

This presents two problems. First, there is a significant de-crease in sensitivity depending on the intensity of S. The absorp-tion signal is no longer $I_0 - I_1$, but is $I_0 - I_1 + S$.

For many elements, particularly those with resonance lines less than 3000 Å, the emission intensity is extremely small, and therefore the error involved in the incorporation of a signal low in intensity S is minimal. But for some elements, such as sodium or potassium, which emit at long wavelengths, the emission signal is large. The net absorption signal $I_0 - I_1 + S$ may be very small or even negative.

The second and more important problem is that S, the inten-sity of emission observed in flame photometry, is subject to numer-ous errors. Therefore it is not possible to calculate or correct for S based on the concentration of the sample (which is generally un-known anyway). As a result, an analytical interference results and inaccurate data are obtained.

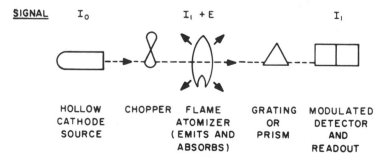

SIGNAL $\quad I_0 \qquad\qquad\qquad I_1 + E \qquad\qquad\qquad I_1$

HOLLOW	CHOPPER	FLAME	GRATING	MODULATED
CATHODE		ATOMIZER	OR	DETECTOR
SOURCE		(EMITS AND	PRISM	AND
		ABSORBS)		READOUT

Figure 2.22. Schematic diagram of a single-beam atomic absorption instru-ment.

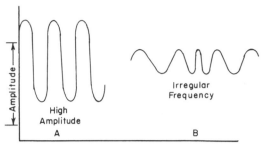

Figure 2.23. Illustration of AC signals. A. Regular high amplitude. B. Irregular frequency.

The problem was ingeniously overcome by Walsh [2], who *modulated* his system. This process involves using an AC or interrupted DC signal from the hollow cathode and tuning the detector to the same frequency. All radiation from the flame is DC or continuous. Since the detector can register only the alternating signal emitted from the hollow cathode, it cannot register the steady signal from the flame. The effect is that the detector reads out only the signal from the radiation source, but does not include the emitted radiation from the flame. This eliminates any interference arising from sample radiation.

There are two methods of providing an AC radiation source. One is to use an intermittent electrical impulse to the hollow cathode, which then radiates an intermittent signal. The second method is to use a rotating quadrant in front of the hollow cathode. The latter alternately (a) reflects the signal and (b) allows the signal to pass down the light path. This provides an interrupted radiation signal, one that alternates between zero and I_0. This device is known as a mechanical chopper. In each case the frequency of chopping is maintained at a very controlled rate.

The amplifier used to amplify the signal from the detector is usually a simple AC amplifier, that is, one that detects any regular intermittent signal. The amplifier measures the amplitude of the signal, whatever the frequency of that signal may be. This is illustrated in Fig. 2.23.

These amplifiers work reasonably well under most conditions and are used in most commercial equipment. One problem with these amplifiers is that they will record a signal that is generated by flame flicker. Because they are able to detect any change in signal intensity, they cannot discriminate between an alternating signal coming from the hollow cathode and flicker in the flame that results in small changes in light signal. The net result is analytical inaccuracy, especially if high-intensity emission from the flame is encountered.

If necessary, this problem can be overcome by using a lock-in amplifier. The lock-in amplifier is tuned to exactly the same frequency as the hollow cathode itself, and does not pick up flicker signals or any other AC radiation that occurs at a different frequency from that of the hollow cathode. The lock-in amplifier provides better accuracy if this is a necessity, but naturally increases the cost of the equipment.

A secondary effect of emission from the flame is fatigue of the detector. Although the intense radiation from the flame does not contribute to the absorption signal, the detector nevertheless is exposed to this radiation and may become fatigued if the radiation is not eliminated. Failure to do this results in unreliable and erratic data.

Correction for Flame Absorption

Molecular species in a flame may absorb at the same wavelength as the atomic absorption line. This introduces a direct source of error.

The error can be eliminated by using one of several techniques. The most common are (a) the use of a H_2 lamp, (b) the use of a nonresonance atomic line at a wavelength close to the absorption line, and (c) the use of the Zeeman effect. These techniques are described in greater detail in Chapter 3.

REFERENCES

1. J. W. Robinson, *Undergraduate Instrumental Analysis*, 2d Ed., Marcel Dekker, New York, 1973.
2. A. Walsh, *Spectrochim. Acta*, 7, 108 (1955).
3. W. G. Jones and A. Walsh, *Spectrochim. Acta*, 16, 249 (1960).

Chapter 3

ANALYTICAL PARAMETERS

3.1 CHOICE OF ABSORPTION LINE

Atomic theory tells us that the electrons in all atoms are in well-defined orbitals. For example, in uranium the electron shells with principal quantum number 1 through 6 are all filled and the shell with principal quantum number 7 is partially filled. Numerous orbitals are available in each shell. These are the s, p, d orbitals, etc. In the filled shells, each orbital accommodates an electron.

Atoms with low atomic numbers are made up of a similar low number of electrons. In the unexcited atom, these electrons reside in the orbitals with the lowest energy level. However, each of the upper empty orbitals is available to accommodate an electron. During excitation the electron with the highest energy (valence electron) moves from its normal low-energy orbital to an orbital with a higher energy. This orbital may be in the same shell or in a higher shell, inasmuch as each orbital is available to accommodate an electron.

In atomic sodium, electrons fill the shells with quantum numbers 1 and 2 and one electron is in the shell with quantum number

3. When sodium is in the ground state, this will be the orbital with the lowest energy, that is, 3s. If we excite sodium, the 3s electron can move to an orbital with higher energy. The energy level next to the 3s level is the 3p energy level, hence it is possible for an electron to go from a 3s to a 3p orbital. It is also possible for the 3s electron to go into orbitals of even higher energy, such as 4p orbitals, 4d orbitals, 5p, 5d, etc.

When the valence electron of sodium is in the 3s orbital, its lowest energy state, the sodium is said to be in the *ground state.* When the electron is in any orbital with higher energy, the sodium is said to be in an *excited energy state,* or we can say that we have *excited sodium.*

An atom can become excited by absorbing energy of the correct wavelength. The wavelengths of the energies involved are well known and follows standard physical laws as represented by Eq. (1.1),

$$E = h\nu \tag{3.1}$$

where

  E = energy difference between the lower energy state and the higher energy state
  h = Planck's constant
  ν = frequency of the radiation

When radiation energy is absorbed, the atom becomes excited. If we use a prism monochromator to disperse the radiation falling on the atoms, the absorption spectrum appears as a narrow line as opposed to a wide band. If the transition is between the ground state and the lowest excited state, then it is said that the absorption is the *resonance line.*

In theory, transition from any excited state to a higher excited state can occur following absorption. In practice, the transitions are not common. Two principal limitations restrict the number of transitions.

First is the limitation of population. Only those atoms that are in a given energy state can absorb suitable radiation to excite it from that state to a higher excited energy state. In practice, at

temperatures encountered in flames the vast majority of atoms
exist in the ground state and the electrons are in the lowest energy
state. Hence, only transitions that involve the ground state are
meaningful. The upper excited states are so sparsely populated
that there are too few atoms available for significant percentage
absorption of the radiation falling on them.

The second limitation is that the energy of the absorbed radiation must be sufficient to cause excitation. In sodium the transition
between the 3s orbital and a 3p orbital can be achieved by absorbing radiation at 5890 or 5895 Å (the sodium D doublet).
Similar absorption of radiation at 3303 Å will cause sodium to be
excited from the 3s ground state to the 5p excited state orbital.
Transitions between the 3s orbital and orbitals with principal
quantum number 6 require a lot of energy. In the case of sodium,
such absorption is quite feasible. But with many elements, such as
zinc, more energy is required than is available from ultraviolet

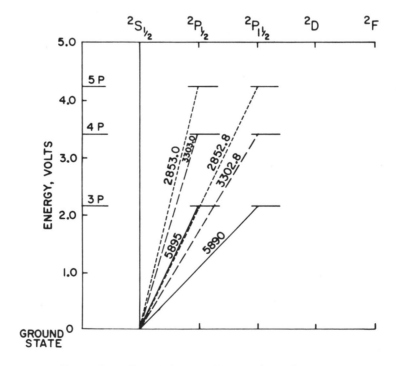

Figure 3.1. Partial Grotian diagram for sodium.

radiation. Consequently, although absorption is feasible, the radiation energy required is so high that it is outside the spectral range of the instrument. In fact these absorption lines would exist in the vacuum ultraviolet region of the spectrum, which cannot be handled by conventional equipment.

The energy levels of sodium are illustrated in a Grotian diagram such as that shown in Fig. 3.1. For the sake of clarity, many of the upper-state transitions are omitted. The Grotian diagram indicates the energy levels of the 3s, 3p, 4p, 4d, etc., orbitals. Such Grotian diagrams are available for most elements in the periodic table.

The most important conclusion is that in atomic absorption spectroscopy the only transitions that can be used are between the ground state and an excited state. Absorption is more intense between the ground state and the first excited state, but absorption is possible between the ground state and other higher excited states.

For the nonmetallic elements such as the halides, the ground-state absorption lines require so much energy that the absorption is in the vacuum ultraviolet and cannot be measured with conventional equipment. Therefore, it is not possible to detect nonmetallic elements by conventional atomic absorption spectroscopy.

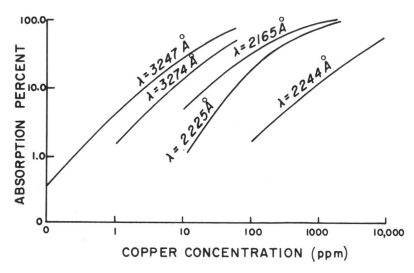

Figure 3.2. Calibration curves for various copper absorption lines.

For some elements, particularly the transition metals, several absorption lines are available for use in atomic absorption spectroscopy. The detection of very small quantities requires the use of the resonance line that absorbs most strongly. However, for increased concentrations absorption lines that absorb less strongly may be used. The ability to select different absorption lines greatly increases the analytical range of the method.

An illustration is the calibration data for copper using several different absorption lines as shown in Fig. 3.2. The analytical range of the method is extended to 10,000 ppm by simply changing the absorption line.

In summary, then, the choice of the correct absorption line depends on the analytical sensitivity required. For high sensitivity, the resonance line must be used; for increased concentrations, other absorption lines can be used.

It should also be emphasized that the useful absorption lines originate in the ground state. In emission spectrography, emission lines occur as a result of an electron dropping from an excited state to a lower excited state or to the ground state. The most persistent lines are those originating from transitions between low excited states and ground states. These are always the most intense lines, and the last lines to disappear as metal concentration decreases. In practice, the strongest absorption lines are frequently the most intense emission lines.

3.2 DEGREE OF ABSORPTION

The ultimate objective of an analytical determination is to obtain accurate and precise results. In order to do this, it is vital that we understand clearly what is being measured and the effect of variables on this measurement.

The degree of absorption is given by Eq. (1.4),

$$\int_0^\infty K\nu \; d\nu = \frac{\pi e_2}{mc_2} Nf \tag{3.2}$$

where

$\int_0^\infty K\nu \, d\nu$ = total amount of light absorbed at frequency ν

$\dfrac{\pi e_2}{mc_2}$ = constants

N = total number of absorbing atoms in the light path
f = oscillator strength

The total degree of light absorbed, therefore, depends on a number of constants times N, the number of atoms in the light path, and f, the oscillator strength. The oscillator strength depends on the probability of the transition between the ground state and the higher excited state. The greater the probability, the greater the oscillator strength. This is a physical property of the atom and does not vary under normal circumstances. We can see, therefore, that the total degree of absorption depends on the product of a number of constants times the number of atoms in the light path.

The number of atoms formed is equal to the number of atoms of the element in the original sample times the efficiency of atomization. This number is modified by the fact that the free atoms are continuously lost as they react in the atomizer to become oxides or other chemical entities. The number of free atoms N is a dynamic equilibrium between the number created and the number lost:

$$\text{sample molecules} \rightarrow \text{free atoms} \rightleftharpoons \text{metal oxides} \qquad (3.3)$$

In order to get reproducible analytical data, it is vital to control both the efficiency of producing atoms and the rate at which the atoms are lost. When this control is effective, the number of free atoms in the system remains constant for a given concentration of sample.

Any variable that affects the efficiency of atomization or the rate of loss of free atoms after atomization directly affects N, which in turn controls the absorption measurement and, therefore, directly affects the analytical data. For reproducible results it is necessary to control these variables, which depend to a large degree on the atomizer being used. In this chapter we shall consider only the flame atomizers, inasmuch as they are by far the most important atomizers used and the most generally available commercially.

3.3 FLAME ATOMIZERS

The sample is introduced as a liquid into the base of the flame of a flame atomizer. Here it is nebulized into small drops, which are evaporated and then atomized as they proceed through the flame itself. The free atoms thus formed find themselves in the hostile environment of a flame and are usually quickly oxidized to some other chemical form. It has been shown on numerous occasions that the efficiency of producing atoms by flames is very low. It is not uncommon to reduce only one atom to the atomic state for every 100,000 chemically combined atoms that are introduced into the flame.

Two main types of burners have been used extensively in atomic absorption spectroscopy. These are the total-consumption burner and the Lundegardh burner. Each burner is deceptively easy to operate and generates analytical data. However, it is only under very carefully controlled conditions that the data are reliable. Total-consumption burners and Lundegardh burners are illustrated in Figs. 2.20 and 2.21.

For the sake of simplicity, we shall discuss the process involved in the total-consumption burner. The same process occurs to a lesser or greater extent in the Lundegardh burner.

Atom Population Profile in the Flame

The normal height of the flame from base to tip of a total-consumption burner is about 3 in. If the atomic absorption is measured at different heights in the flame, it is found to vary considerably. The relationship between height in the flame and atomic absorption signal is called the *flame profile*. Profiles of three different elements are shown in Fig. 3.3, in which it can be seen that the absorption is very low at the base of the flame, increases rapidly to a maximum at the reaction zone, and then decreases slowly as the top of the flame is reached.

The degree of absorption is a direct measure of N, the number of atoms in the light path, and therefore the flame profile is a measure of the atomic population in various parts of the flame. It should be noted that the emission maximum takes place at a low

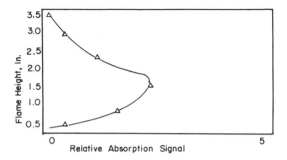

Figure 3.3.   Flame profile for nickel 3414-Å line: relative absorption vs. flame height.

part of the flame, presumably because the lifetime of an excited atom is quite short (10^{-6} to 10^{-7} sec). But free atoms have an infinite lifetime provided that they do not react chemically or become absorbed on the surface. Hence, an accumulation of free atoms occurs as we move into the upper regions of the flame.

Generally, the sample enters the flame base as a liquid droplet, and the metal of interest leaves it in the form of an oxide. Numerous steps are involved in between. These are illustrated in Table 3.1. It will be appreciated, of course, that many of the larger droplets never evaporate completely and therefore pass through the flame virtually unaffected, taking with them atoms of the element being determined. Also, quite often the residue left behind after evaporation passes through the flame without further decomposition to free atoms. In each case the atoms contained in the unevaporated droplets or the undecomposed residues pass through the flame and never reach the atomic state. Consequently, they never contribute to the absorption signal.

Factors Affecting Atomization

There are two major segments of the flame absorption profile: (a) At the base of the flame, where absorption signal increases, atoms are formed and accumulated. (b) In the upper regions of the flame, where the signal decreases, the atoms are oxidized and no longer absorb strongly at the atomic absorption lines. Various factors affect the formation and loss of atoms from the system. The more important ones are as follows.

Table 3.1

Factors Affecting Flame Profiles

Physical form of sample in flame	Reaction	Factors controlling reaction	Part of flame
Oxide	No reaction or reduction	Stability of metal oxide, flame composition	Outer mantle
Atoms	Accumulation or oxidation	Flame composition, stability of atoms	Reaction zone
Solid particles	Disintegration	Stability of compound, anions, flame temperature, ultraviolet light emitted from the flame	Inner cone
Droplets	Evaporation	Droplet size, solvent, flame temperature feed rate, combustibility	Base

Droplet Size. The sample is aspirated into the base of the flame by the burner. Here it is broken down into small droplets, which pass through the flame. If the droplets are relatively large, their surface area per unit weight is small and evaporation is slow. It is quite possible under these conditions that they will pass through the flame without completely evaporating, and that any metals contained in them are never reduced to the free atomic state and never contribute to the absorption signal. On the other hand, if the drop size is small, then the surface area is relatively large per unit weight and evaporation is rapid. The droplet evaporates, leaving a solid metal residue which is then decomposed by the thermal energy of the flame. Any metal in the residue is available for reduction to the free atomic state and can contribute to the absorption signal. Most total-consumption burners produce large-, intermediate-, and small-sized droplets, and thereby operate at an acceptable but not optimum efficiency. In contrast, the Lundegardh burner produces very small drops or even residual particles which are swept by the gas

flow into the base of the flame. Its efficiency in atomization is
therefore enhanced by this step.

A second important consideration related directly to drop
size is the size of the solid residue remaining after evaporation of
the drop. If the size is large, then the residue will be relatively bulky
and decomposition by the thermal energy of the flame will be slow
and inefficient. This again will cause a loss of sensitivity, because
any metal residue that is not decomposed cannot contribute to the
absorption signal. Large residues can be caused by a high salt content in the sample, or by the evaporation of large drops by a hot
flame.

Another factor affecting the efficiency of decomposition of
the residue is the chemical stability of the solvent involved. For example, we may have two aqueous solutions of aluminum, each with a
concentration of 10 parts per million. One solution might be aluminum chloride and the second solution might be aluminum hydroxide. When the solutions are introduced into the base of the flame,
the solvent evaporates off and we are left with a residue including
aluminum chloride in one case and aluminum hydroxide in the
second case. Even if the drop size is equal, it is still likely that
aluminum chloride will be more easily broken down to produce
free aluminum atoms than aluminum hydroxide. Hence, we would
get a stronger absorption signal from a solution of aluminum chloride than from a solution of aluminum hydroxide. This phenomenon
is at the heart of chemical interference and is a major source of
error in atomic absorption spectroscopy. It will be discussed in
Section 3.11.

Drop size is very dependent on the design of the nebulizer
section of the burner. The three most important factors controlling
nebulizer efficiency are as follows: (a) Maximum sample breakup
(minimum drop size) occurs when the sample is introduced into a
point of maximum gas velocity. The greater the gas velocity, the
lower the particle size. (b) Subjecting the sample to a large velocity
gradient decreases the drop size. An obstruction in the gas carrying
the sample droplets causes a decrease in drop size. (c) An abrupt
change in pressure, velocity, or direction of the gases—such as at a
shock front—will break up the drops to smaller size.

These factors are usually taken into account by commercial
manufacturers of burners. The burner design has already been op-

timized by the manufacturer, and the operator is not called upon to change the design of the burner to be used except under research conditions.

Sample Feed Rate.   Droplet evaporation and residue disintegration require energy, which is provided by the flame in the total-consumption burner. In the Lundegardh burner and the preheat burner, the evaporation takes place in the mixing chamber and relieves the flame of some of this function. Under controlled conditions the flame produces energy at a steady rate. If the sample feed rate is too high, too much of the energy of the flame is used up in nebulizing and disintegrating the sample. The final atomization step is therefore inefficient. Under these conditions the flame is swamped with the sample. On the other hand, if the sample feed rate is too low, the production of neutral atoms is reduced and again the absorption signal is diminished. Between these extremes there is an optimum feed rate, which varies with the burner design and even between different burners of the same design. For reducible results the sample feed rate must be kept constant.

As can be readily imagined, change in the solvent causes a significant change in the droplet size because of viscosity changes in the sample. Some burners are adjustable and corrections can be made for such changes, but this is not usually the case. Whenever possible, the burner should be optimized for use. In all cases the conditions used for making up calibration curves should be the same as those used for analyzing the sample. If a different solvent is used in the preparation of the calibration curve than is used in the sample, the data will not be relevant and inaccurate results will be obtained.

Flame Temperature.   The transformation from a solution to free neutral atoms is achieved by absorbing energy from the flame. The energy of the flame is directly related to its temperature.

The hotter the flame, the more efficient it will be in reducing the sample to free atoms. The efficiency of production is related to drop size and to the reduction of free atoms in the residues formed. With low-temperature flames, small changes in the chemical energy required to bring about reduction to atoms can cause significant changes in efficiency of atomization. But with high-temperature flames the efficiency is much greater, and similar small changes in chemical energy have a reduced effect on atomization efficiency. It should be pointed out, however, that if the amount of energy required to break down two different chemical forms is great, then there is a significant change in atomization efficiency even when hot flames are used.

These energy changes are brought about by the chemical composition of the sample and are the basis of chemical interference. Differences in chemical stability between two chemical forms directly affect the ease of atomization, and therefore the absorption signal. Low-temperature flames are very subject to chemical interference and should be avoided if possible. High-temperature flames are much more free from chemical interference and are to be recommended when available. Unfortunately, interferences are not completely removed by high-temperature flames.

The temperatures of typical flames are shown in Table 3.2. Flames commonly used in atomic absorption spectrocopy include air-acetylene, nitrous oxide-acetylene, oxyacetylene, oxyhydrogen, and propane-oxygen.

Oxide Formation. After metals in the sample have been reduced to neutral atoms, they will stay in this state for varying periods of time. This time period ends if and when the atom reacts with the oxidant or other components of the flame. It should be

Table 3.2

Flame Temperatures of Typical Flames

Fuel	Oxidant	Flame temperature ($^{\circ}$C)
H_2	O_2	2800
H_2	Air	2100
H_2	Ar	1600
Acetylene	O_2	3000
Acetylene	Air	2200
Acetylene	N_2O	3000
Propane	O_2	2800
Propane	Air	1900

pointed out that an oxidizing flame is one in which there is an excess oxidant, and a reducing flame is one in which there is an excess
fuel; in either case oxidation of metal to metal oxides can take
place by reaction with the oxidant. However, oxide formation is
more likely in an oxidizing then in a reducing flame.

The ease with which a metal atom is oxidized depends on the
chemistry of the particular metal. If the oxide is easily formed and
is stable, it will oxidize very easily. The step is controlled by the
equilibrium

$$\text{metal} + \text{oxidant} \xrightarrow{\text{K}_{eq}} \text{metal oxide} \tag{3.4}$$

It can be calculated that K_{eq} at $3000°C$ for silver is about
$10^{-2.7}$ and for magnesium about $10^{1.9}$. Their respective flame profiles are shown in Fig. 3.4 and Table 3.3. It can be seen that there
is a marked difference between the profiles of magnesium and silver. The magnesium absorption goes through a sharp maximum at
the reaction zone of the flame. Optimum sensitivity is obtained
only when the light path passes through this maximum. For reducible quantitative results it is important that the light path always pass
through the same part of the flame, both when calibration data are
obtained and when the samples are being analyzed. It must be
remembered that the total height of the flame may vary somewhat
from day to day. It is therefore important that the absorption maximum be located by actual measurement.

It can be seen that the silver absorption curve does not go
through a maximum. This is because silver oxide is easily decomposed and does not form in the flame. Consequently, an accumulation of neutral silver atoms takes place in the higher parts of the
flame. This type of flame profile is common with the noble metals.

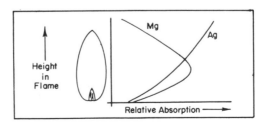

Figure 3.4.   Effect of oxide stability on flame profile. MgO = stable oxide;
Ag_2O = weak oxide.

Table 3.3

Flame Absorption Profile for Silver and Magnesium

Metal	Flame height for maximum absorption	Predicted stability of oxide
Mg	Medium	Medium
Ag	High	High

Variation in the Ratio of Fuel to Oxygen in the Flame. The maximum temperature of a simple flame such as an oxyhydrogen flame is achieved when the ratio of oxygen to hydrogen is the stoichiometric ratio necessary for complete combustion of each component. When there is an excess of oxygen, the flame temperature drops and there is an excess of hot oxygen in the flame plasma. Similarly, when there is an excess of hydrogen, the flame temperature drops and there is an excess of hot hydrogen in the flame. Excess oxygen produces an oxidizing flame, and excess hydrogen produces a reducing flame.

Metal oxides form more rapidly in oxidizing flames than in reducing flames. Although each flame may have the same temperature and may produce the same number of free atoms from a given sample, the rate of loss of free atoms will vary depending on whether it is an oxidizing or reducing flame.

It must also be remembered that in such flames there is an abundance of hot water molecules, of OH, $O_2 H$ and other reactive chemical entities. These are all capable of oxidizing the neutral atoms even in reducing flames.

We can see that the oxidant-fuel ratio directly affects the flame temperature and the lifetime of neutral atoms in the flame. It therefore directly affects the absorption signal that will be obtained from a given sample. For reproducible results the ratio of oxidant to fuel in the flame must be kept constant. For most metals these ratios have been worked out and are published in the literature. Recommended flow rates are also suggested by burner manufacturers.

Nitrous Oxide-Acetylene Flames

Independently and simultaneously, Amos [1] and Willis [2] investigated nitrous oxide-acetylene flames. This flame is very hot and requires the use of a laminar flow burner to assure maximum atomization. The use of nitrous oxide instead of oxygen as the oxidant greatly reduces the probability of a metal atom becoming oxidized in the flame. Results have shown that they are particularly useful for atomizing metals that form refractory oxides. In other flames these metals are reduced only with great difficulty. A partial list of such metals and their sensitivities is given in Table 3.4.

The nitrous oxide-acetylene flame has been shown to have applications for other metals as well as those that are difficult to reduce. It is therefore universal in application and is used extensively for atomic absorption spectroscopy.

Table 3.4

Metals with Refractory Oxides Reduced in N_2O-C_2H_2 Flames

Metal	Sensitivity (ppm)
Al	1.0
Be	0.1
Sc	1.0
W	1.0
U	100.0
V	1.0
Zr	50.0

Conclusions

Flame atomizers have been demonstrated to be a very useful and reliable tool in atomic absorption spectroscopy. They are sufficiently efficient to provide an analytical technique to carry out analyses of the part per million level or lower, and this satisfies most needs. Their efficiency in atomization is known to be low, and there is usually a loss of metal from a system by oxidation. However, the method has been very reliable for quantitative analysis and is widely used.

Much higher sensitivity is being achieved by using other atomizers. These will be discussed in Chapter 7.

3.4 CATIONS

In general the presence of several cations in the sample has no effect whatever on the absorption signal of the metal of interest. It has been noted, however, that if the concentration of one interfering cation is very high and its resonance line is very close to the resonance line of the element being determined, then some apparent interference can take place because of absorption of the "wings" of the resonance line of the concentrated metal. This case is unusual, and in the majority of instances is of academic interest only.

It can be shown that even different isotopes of the same metal do not absorb at the same resonance line. This has been demonstrated for lithium-6 and lithium-7 and for uranium-235 and uranium-238. Isotope analysis can therefore be achieved by using a lithium-6 hollow cathode lamp as the radiation source and determining the amount of lithium-6 in the sample. No absorption by lithium-7 takes place.

Mutual interference has been noted in the case of magnesium and aluminum. When these two metals are present in the same solution and are aspirated into the flame, there is a decrease in the absorption signal compared to that which would be obtained if only one element were present. This might be explained by the formation of an intermetallic compound, which is not decomposed in the flame. A more likely explanation is the formation of metal oxide salts such as magnesium aluminate. These compounds are difficult to decompose, and their formation would result in a decrease in the formation of both magnesium and aluminum atoms in the flame.

If the sample is known to contain varying amounts of these elements, the problem can be overcome somewhat by using the standard method of addition for calibrating the system.

Although these cases of cationic interference have been observed, there are very few other cases reported of this type of interference. In general, therefore, atomic absorption spectroscopy is free from cationic interference.

3.5 ANIONS

Normally in an aqueous solution the metal is in the form of a cation and is dissolved with a corresponding anion. When the solution is aspirated into the flame, droplets are formed which evaporate and leave the residue. The residue contains the metal cation and anion combined to form a molecule. In order to produce neutral atoms from this molecule, the metal-anion bond must first be broken. The strength of this bond determines how easily the step is taken. A strong metal-anion bond is difficult to break and, consequently, for a given flame system production of free metal atoms is less efficient. If, however, the metal-anion is a weak bond, then it will more easily be broken and production of the free atoms will be more easily accomplished. In the latter case, the increased number of free atoms will give a greater atomic absorption signal even though the concentration of the metal in the solution is constant.

If the metal is combined with an organic anion, the situation is somewhat modified but is still similar. The organic anion will burn in the flame, and the free metal atom is liberated quite easily. This results in a greater absorption signal and corresponding increase in sensitivity. It can be seen, therefore, that the absorption signal will depend on the molecular form of the element being determined. An illustration of this chemical effect is shown in Table 3.5. The effect is to change the metal compound form in the absorption signal. This effect can be observed by simply adding different anions to a solution, particularly if these anions are in a large excess compared to the metal being analyzed. For example, a solution of lead nitrate may be made up such that its concentration is one part per million; but if phosphate ion is added to the solution at a concentration of 100 parts per million, when the solution goes

Table 3.5

The Effect on the Absorption Signal of Changing
the Metal Compound Form

Compound	Solvent	Chromium absorption signal, $I_0 - I_1 \ (I_0 = 100)$
Chromic nitrate	Ethanol	20.5
Sodium chromate	Ethanol	18.2
Chromium naphthenate	Ethanol	26.3

to the normal atomization step there is a high probability that the
lead will form lead phosphate rather than lead nitrate in the residue.
In order to achieve atomization it is necessary to break the lead
phosphate bond rather than the lead nitrate bond. A summary of
this effect is shown in Table 3.6.

The effect can be greatly reduced or even eliminated by adding a strong organic complexing agent. For example, by adding
EDTA in the example shown in Table 3.6, the lead will form the
lead EDTA complex because the latter complex is more stable than
the simple ionic solutions. Under these circumstances the lead will
exist in the residue state preferably as the EDTA complex. Therefore, change in signal due to the presence of difference anions will
be eliminated. It can be seen that the signal remains about the same
irrespective of the predominant anion present in the solution.

This is a useful way to eliminate anion effect when samples of
unknown composition are being analyzed.

Table 3.6

Chemical Interference and Its Elimination

Conditions	
Sample	Lead nitrate solution, 1.0 ppm
Wavelength	2170 Å
Flame	Oxyhydrogen 30 liters/min:50 liters/min
Burner	Beckman total-consumption
Aqueous	Solution

Interfering anion (100 ppm)	Absorption (%)	
	Lead nitrate solution, plus interfering anion	Lead solution plus anion plus EDTA (1%)
None	23.0	23.0
PO^{3-}	19.9	22.8
Cl^-	22.3	22.3
CO_3^{2-}	12.8	23.0
I^-	13.7	23.5
SO_4^{2-}	22.0	23.0
F^-	21.3	22.9

3.6 THE EFFECT OF DIFFERENT SOLVENTS

The solvent has several direct and indirect effects on the absorption signal. These effects are caused by: (a) spectral emission by the solvent or combustion products of the solvent when introduced into the flame, (b) spectral absorption by the solvent or its combustion products in the flame, (c) the effect of the solvent on the efficiency of producing neutral atoms from the sample, (d) the change in sample feed rate with solvents with different viscosities, and (e) the effect of surface tension on drop size when the sample is introduced into the flame.

Spectral Emission from the Solvent and Solvent Combustion Products in the Flame

If we examine the emission signal from a simple oxyhydrogen or oxyacetylene flame, we shall observe several strong emission bands. If we then aspirate aqueous or organic solvents into the flame, we shall find that more emission bands will be present. The emission from such flame systems has been studied extensively by Gordon [3]. The most common emission bands found are those of the OH radicals, which are formed either when water is introduced into a flame or organic compounds containing hydrogen are burned—thus producing water in the flame. The mechanism of production of these emission bands will not be discussed in this book; they have been adequately described by Gordon. The emission does not represent a problem because it can be eliminated as an interferent by modulating the system. As in atomic emission, any molecular emission signal is DC and does not register when the system is modulated. Trouble can arise if the emission signal is very intense, as in the case of an oxyacetylene flame. When intense radiation falls on the detector, the latter becomes fatigued and erratic, thus producing unreliable data.

The Absorption of Spectral Energy by the Flame and Solvents

Simple oxyhydrogen flames are absorbed over an extensive spectral range. The typical absorption curve is shown in Fig. 3.5.

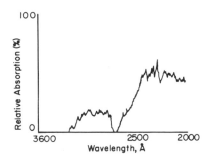

Figure 3.5.   Absorption spectrum of oxyhydrogen flame.

The absorption is probably due to OH moieties present in the
flames. It can be seen that it is quite intense over the region 3000–
3200 Å.

Similar absorption is observed when other organic solvents
are introduced into the flame. The degree of absorption is greater
when solvents that contain halides such as CCl_4 are used. This is
particularly true at wavelengths less than 3000 Å. This is molecular
absorption, and although it is not strong it is still not negligible and
can cause serious error if corrections are not made. Inasmuch as
these are molecular absorption bands, they are not continuous, but
are a series of closely adjacent absorption lines. Table 3.7 gives a
list of some spectral lines that are absorbed and a second list of
lines in the same region that are not absorbed.

The molecular absorption varies significantly with the part of
the flame being examined. It is important, therefore, that all absorption measurements be taken at the same part of the flame in
order for the correction to be valid. There are two ways to correct
for this source of error.

The first method is to measure the absorption of the metal
resonance line by the flame and the pure solvent alone. This gives
a direct measure of the absorption by the flame system at the metal
absorption line. The sample is then analyzed measuring the total
absorption of sample element plus the flame system. The difference
between the two measurements is the absorption due to the sample
element being measured.

The second method of correction for molecular absorption is
to use a hydrogen lamp in conjunction with the hollow cathode
lamp. The principle is as follows. Let us suppose that we wish to

Table 3.7

Nonabsorbed Lines for Background Correction

Element	Wavelength of hollow cathode line absorbed (Å)	Wavelength of hollow cathode line not absorbed (Å)
Aluminum	3092	3070
Antimony	2176	2179
Cadmium	2288	2276
Chromium	3579	3520
Cobalt	2407	2388
Copper	3247	3234
Indium	3040	3057
Iron	2483	2472
Lead	2833	2820
Magnesium	2852	2817
Molybdenum	3133	3234
Nickel	2320	2316
Platinum	2659	2702
Tin	2863	2839
Vanadium	3184	3125
Zinc	2139	2125

correct for molecular absorption while measuring the atomic absorption of magnesium. The system would be tuned to the wavelength of the magnesium resonance line, that is, 2833 Å. Here the absorption of the sample would be measured. This signal would be composed of the atomic absorption by magnesium and the molecular absorption by any molecular species that absorbed in this spectral region. Let us now replace the hollow cathode with a hydrogen lamp, but leave the monochromator at the same wavelength setting. The molecular absorption over the 2-Å spectral range will remain as before. But the atomic absorption will remove radiation only from a narrow line (0.01 Å) from the total bandwidth of 2 Å. The total light absorbed from the hydrogen lamp will be the molecular absorption over the entire bandwidth plus the atomic absorption over an extremely narrow bandwidth. Even if the atomic absorption succeeded in removing all the radiation from a bandwidth of 0.01 Å, the decrease in signal would only be 0.5%. Absorption of the hydrogen lamp by the magnesium is very small and can be ignored. Hence, the molecular absorption signal can be measured by simply measuring the absorption from a hydrogen lamp at the same wavelength as that in which the resonance line occurs. This is illustrated in Fig. 3.6.

Correction for Background Absorption Using the Zeeman Effect [4, 5]. When an element such as mercury emits or absorbs radiation at a resonance line (e.g., 2533 Å for Hg), the "absorption line" actually consists of several absorption lines which are very close together and are characteristic of each isotope of mercury. If we direct our attention to one isotope, such as ^{204}Hg, it emits and absorbs at only one characteristic wavelength.

However, if we place the Hg lamp source in a magnetic field, the single line is split into two lines by the Zeeman effect. The shift is about $\pm 10^{-4}$ Å. If the magnetic field is strong enough, the original unshifted line disappears; otherwise, the shifted and unshifted lines are emitted. This is illustrated in Fig. 3.7.

Typical magnetic fields are between 7 and 15 kgauss. The magnetic field has two important effects on the emission line. First, it causes splitting and a shift in wavelength. Second, it changes the

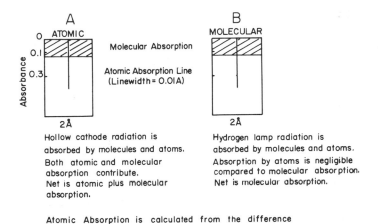

Figure 3.6.   Correction for molecular absorption.

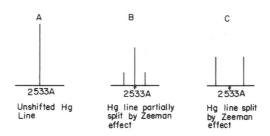

Emission spectrum of Hg 2533A (A) unshifted, and (B)
and (C) after putting the source in a magnetic field.

Figure 3.7. Zeeman effect causing shifting of emission lines.

polarization of the light so that the radiation that is split (Fig. 3.7B)
is polarized perpendicular to the radiation that is not split (Fig.
3.7A).

The unsplit radiation (Fig. 3.7A) is at the natural wavelength
of the Hg resonance line and is absorbed by the Hg atoms of the
sample (and the molecular absorption). The Zeeman-shifted radiation (Fig. 3.7B) is not at the resonance wavelength and is not absorbed by the Hg atoms. However, this radiation is absorbed by the
background molecular absorption. By measuring each absorption
signal it is possible to correct for the molecular absorption signal.

This has been achieved in two ways. First, the applied magnetic field may be alternating, and by turning the amplifier to this
frequency it is possible to discriminate between the split and unsplit radiation. In practice this is difficult.

A second method is to take advantage of the fact that the
split and unsplit light is polarized differently. Fused quartz under
pressure is birefringent, that is, light is diffracted differently when
its axis of polarization is different. The quartz may be stressed at a
controlled frequency and the two beams of light discriminated. By
suitable amplification the background absorption—measured by the
Zeeman-shifted radiation—can be automatically corrected for.

The advantage of this technique is that only one source and
only one detector are used. The background correction is made at a
wavelength very close to the resonance line and is therefore accurate. Finally, the system has the advantage of being compact and
relatively easy to operate.

A major difficulty with the technique is that the magnetic
field used to generate Zeeman splitting also interacts with the ions

in the hollow cathode. This causes the emission from the hollow cathode to be noisy, which in turn introduces imprecision into the procedure [4].

Summary.   Molecular absorption by solvents is usually not very great, but can be a significant source of error particularly when aromatic or halide solvents are used. Correction must be made in order to obtain reproducible analytical data. This can be done by simply measuring the absorption of the resonance line by the solvent plus flame, by measuring total molecular absorption with a hydrogen lamp, or by taking advantage of the Zeeman Effect.

The Effect of Solvents on the Efficiency of Producing Neutral Atoms

It has been observed on numerous occasions in both flame photometry and atomic absorption spectroscopy that an enhancement of the signal takes place if an organic solvent is used instead of an aqueous solvent. The enhancement of the emission signal and the absorption signal are not equal to each other, but it can be stated unequivocally that in each case a significant enhancement is observed. This can be shown as in Table 3.8. The most significant difference between an organic and a aqueous solvent is the combustibility of the sample. If an organic solvent is introduced into a flame it burns readily, giving off thermal energy—that is, it is an exothermic reaction. On the other hand, if an aqueous sample is

Table 3.8

Effect of Solvent on Nickel Absorption and Emission Signals[a]

Solvent[b]	Emission	Absorption
Water	4	4
Acetone	18	144
n-Pentane	19	66
Xylene	17	30

[a]From Fuwa and Vallee [6].
[b]Experimental conditions: nickel concentration = 10 ppm, absorbtion wavelength = 3434 Å, total-consumption burner.

introduced into a flame, it requires energy to cause it to evaporate. The latter is an endothermic reaction, which proceeds less efficiently than the organic exothermic reaction.

The second factor is the chemical form of the sample in the solvent. If the solvent is aqueous, then usually the sample is ionic. On the other hand, if the solvent is organic, then generally the metal exists as a metal-organic compound. When the solvent is evaporated off completely, we are left with a residue containing the metal of interest. If it is a metal ion salt, then energy is again required to break it down to the free neutral state. However, if it is a metal-organic compound, then the organic addend is combustible and will react in the flame and release the free neutral atoms more readily than in the case of the metal salt.

The process of combustion of the solvent in the flame can be illustrated in Table 3.9. If we insert numerical values for the number of atoms of each step in the process, we can see that organic solvents cause an increase in the number of excited and unexcited atoms produced by this system.

It should be remembered that in atomic absorption the signal depends on the number of free atoms in the ground state. This number is very close to the total number of atoms in the system. However, in flame emission the excitation signal depends on the number of atoms in the particular excited state from which radiation occurs. This number will be very dependent on other factors

Table 3.9

Solvent Combustion Mechanism

Combustion of an Aqueous Solvent			
	step 1	step 2	step 3
Ions in water	→ Evaporation leaving a hydrated residue	→ Residue (dehydrated atoms formed)	→ Excited atoms (emission) Neutral atoms (absorption)

Combustion of an Organic Solvent			
	step 1	step 2	step 3
Metals in organic solvent	→ Solvent burns	→ Organic addend burns	→ Excited atoms (emission) Neutral atoms (absorption)

besides the efficiency of producing atoms. Hence, although we shall observe an increase in both flame emission and absorption, there will not be a direct relationship between the two signals.

It should be added that there is significant evidence that the enhancement effect is not caused by increase in flame temperature when an organic solvent is used. Such temperature increase is bound to take place and some enhancement will occur, but it will not be in the same order of magnitude as that observed in practice.

Change in Sample Feed Rate When Solvents Having Different Viscosities Are Used

When a fluid solvent is used, the sample is aspirated into the base of the flame much more rapidly than when a viscous sample is used. This directly affects the sample feed rate, and therefore affects the absorption signal. Commercial burners are designed to introduce the sample into the base of the flame. Usually it is assumed that an aqueous sample will be used. The use of organic solvents changes the characteristics considerably. The increased flow rate may cause overloading of the sample, or a decreased flow rate due to a viscous sample may cause the sample injection rate to be significantly reduced.

A mechanical burner has been designed that injected the sample at a fixed flow rate independent of its viscosity. It was shown that using this burner all organic solvents gave approximately the same signal and that there was virtually no difference between highly viscous samples and highly fluid solvents. The conclusion that could be drawn from these data was that the viscosity per se did not affect the signal significantly, but that it did affect the flow rate into the sample, and hence the absorption signal if no correction was made.

Effect of Surface Tension

If a sample has a high surface tension, then large drops of the sample are quite stable. If, on the other hand, the surface tension is low, then only small drops are stable. It can be readily understood that the drop size of the solvent after introduction into the base of the burner will depend very much on the surface tension of the solvent being used. This problem was studied using the mechanical burner mentioned previously. In this instance the sample was broken up by jet force to produce drops of approximately the same size in each case independent of the viscosity of the solvent. The data showed that the absorption signal was independent of the surface tension when the drop size was equalized. This showed that the surface tension per se was not an important attribute to the absorption signal, but that it did affect the drop size and therefore did affect the absorption signal obtained in practice.

Data obtained using the mechanical flow rate burner together with data obtained from a simple Beckman total-consumption

Table 3.10

Absorption and Emission of Nickel 3414-Å
Resonance Line in Various Solvents
(10 ppm N_1)

| Solvent | Force-feed burner | | Beckman aspiration burner | | |
	Absorption, $I_0 - I_1$	Emission	Total absorption	Absorption corrected for feed rate	Viscosity at 20°C (1000n)
Acetone	14.5	33	15	7.0	3.3
n-Heptane	15.5	40	—	—	4.2
Ethyl acetate	13.0	37	12	7.5	4.5
Methyl alcohol	15	29	9	5.8	5.9
Benzene	16.0	42	2	1.9	6.5
Toluene	15.0	38	16	15.5	7.7
Methyl cyclohexane	16.0	39	20	14.0	7.7
Amyl acetate	13.5	36	—	—	8.0
Cyclohexane	15.0	35	8	7.3	9.3
Carbon tetrachloride	8.0	5	5	4.0	9.6
Ethyl alcohol	15.0	31	6	3.7	12.0
Nitrobenzene	13.5	34	2	1.0	19.8
Isopropanol	15	33	—	—	22.5
Varsol	14.5	30	5	9.3	—
Water	4	4	3	3.0	10.0

burner are illustrated in Table 3.10. The data indicate that the absorption signal was virtually independent of the viscosity of the sample when the force-feed burner was used. The use of noncombustible solvents such as water and carbon tetrachloride produced a significant change, indicating that combustibility was a factor in producing free atoms. With the total-consumption burner there was a tendency for the absorption signal to decrease as viscosity increased. This tendency was less apparent when a correction for simple feed rate was made. For the samples tested the results were perhaps more noteworthy for the lack of correlation between absorption and viscosity.

Conclusions

It can be concluded that the solvent has a very dramatic effect on both emission and absorption signals. The primary effect is on the efficiency of producing free atoms, mainly because of increased combustibility of the solvent. However, secondary effects caused by viscosity and surface tension were also noted, particularly when aspiration burners were used.

The most important step to be taken to correct for this problem is to be sure that the calibration solvent in the sample and that used in the calibration curves are the same.

3.7 THE EFFECT OF RESOLUTION ON SENSITIVITY

The function of the monochromator and the slit system is to separate the resonance line from the other radiation that emerges from the hollow cathode. The radiation emitted from the cathode consists of the atomic spectra of the metal used in making up the cathode and the spectra of the filler gas. These spectra are made up of a number of characteristic narrow lines. In addition to these intense lines, there is a very low-intensity background radiation, which covers the entire spectral range of the ultraviolet.

In the ideal case only the absorbable resonance line would reach the detector; all other radiation would be eliminated by the slit system. In practice, this is seldom completely possible, but with a suitable monochromator-slit system much of the unwanted radiation can be prevented from reaching the detector.

The first step in achieving this state is to use a good monochromator. The latter disperses the light according to wavelength and separates all the various lines from each other. However, it does not eliminate the background radiation. The best that can be achieved with a good monochromator, therefore, is to provide enough resolution of the line spectra to resolve the resonance line from the other lines emitted by the cathode.

3.8 THE EFFECT OF CHANGING THE SLIT WIDTH

The monochromator slit system is shown in Fig. 2.15. Note that the monochromator provides dispersion of the spectrum and the slits selects the desirable line in the system.

The slit system is composed of an entrance slit and an exit slit. The entrance slit eliminates all stray radiation and prevents it from continuing down the light path. The exit slit prevents radiation from the hollow cathode with incorrect wavelengths from continuing down the light path. The wider apart are the slit jaws, the greater will be the wavelength range permitted to pass through the exit slit. Conversely, the narrower the slits, the narrower will be the spectral range that passes through the slit. The physical distance between the two slits is called the *mechanical slit width*; the wavelength range that is allowed to pass through the exit slit is called the *spectral slit width*.

In practice, it is desirable to allow the resonance line, but no other line, to pass through the spectral slit. As they emerge from the hollow cathode the lines are extremely small, that is, of the order of about 0.01 Å or less. This is much smaller than can be achieved by narrowing down the mechanical slit width. Sometimes it is not possible to close a mechanical slit sufficiently to prevent a second line from reaching the detector. Under these circumstances there is a loss in sensitivity in the analytical system. This is illustrated in Table 3.11. The lines that reach the detector consist of an absorbable line and a nonabsorbable line. Therefore it is not possible to absorb radiation completely from the hollow cathode.

Table 3.11

Sensitivity Loss Caused by Neighboring Unabsorbed Line

Intensity of initial signal	Intensity of signal after absorption	Sensitivity
Resonance line only, I_0	I_1	$\dfrac{I_0 - I_1}{I_0} = 1\%$
Resonance line and second unabsorbed line, $I_0 + I_0'$	$(I_0 - I_1) + I_0'$	$\dfrac{(I_0 - I_1) + I_0'}{I_0 + I_0'} = 1\%$

Under these circumstances, sensitivity is decreased and the calibration curve is much flatter than desirable. If it is not possible to resolve two lines that are very close together, the problem can sometimes be overcome if the unabsorbed line originates with the filler gas rather than the metal. In this case a different filler gas can be used, for instance, helium. But if the unabsorbed line is from the element being analyzed, then the only other alternative is to move to a different resonance line which, hopefully, will not be close to an unabsorbable line from the spectrum.

Changes in Sensitivity with Unabsorbed Background Radiation

Even if there is no emitted line from the hollow cathode close to the resonance line, there is always a small amount of background radiation that cannot be resolved from the resonance line. The effect of this unabsorbed radiation is exactly the same as that of an unabsorbed emission line. In general, the intensity is low and no serious error is involved. However, as a matter of principle, it is better to operate at as narrow a spectral slit width as is possible. An illustration of the effect of increasing slit width on sensitivity is shown in Table 3.12. It can be seen that there is rapid loss in sensitivity as the mechanical slit width, and hence the spectral slit width, is increased.

Table 3.12

Effect of Spectral Slit Width on Sensitivity [a, b]

Mechanical slit width, mm	Spectral slit width, Å (½-wave)	Sensitivity, ppm (1% absorption)
0.015	3.2 (measured)	5
0.05	3.2 (measured)	5
0.1	4.7 (measured)	6
0.4	7.5 (measured)	7
0.6	11.5 (measured)	10
1.0	19.5 (calculated)	50
1.5	29.5 (calculated)	100

[a]From Menzies [7].
[b]Spectral line used, Fe 3719 Å.

In commercial equipment fixed slits are often used. In this case the manufacturers have built in a fixed mechanical slit width that is considered to be acceptable for most purposes. For some cases, particularly in research work, a manual slit is desirable where the slit width can be increased or decreased at will depending on the particular sample being analyzed or studied.

There is no hard and fast rule concerning dispersion and slit width that applies to all elements. The spectrum from each element and filler gas must be examined separately. If there is an unabsorbable line in the immediate vicinity of the resonance line, then some other precaution must be taken—such as changing the filler gas or moving to another resonance line.

It is always good policy to operate with a maximum of dispersion and a minimum of spectral slit width.

3.9 EXCITATION INTERFERENCE

In flame photometry, when the sample is introduced into the flame the metal atoms become excited. Suppose that the sample contains two metals, A and B. It is possible that excited metal atoms A transfer energy to unexcited metal atoms B. The intensity of emission from metal B becomes greater. This is called excitation interference, and is particularly important with alkaline metal analyses.

The effect is important in emission spectrography and flame photometry because it directly affects the number of excited atoms of the elements present in the system. However, the effect is much less important in atomic absorption spectroscopy because it is the unexcited atoms that provide the absorption signal. Excitation interference processes have very little influence on the total number of unexcited atoms present in the system, and therefore interference from this phenomenon is of small importance in atomic absorption spectroscopy.

3.10 RADIATION INTERFERENCE

In flame photometry the intensity of emission from the sample element is measured at some particular wavelength. If another element present in the sample emits at the same wavelength, there is a direct positive interference called radiation interference. The source of this interference is any element or compound that emits at the wavelength being used for measurement.

There are four principal sources of radiation interference. These are broad background emission from the flame, background emission from the solvent introduced into the flame, line emission or molecular emission from other elements in the sample, and finally, the emission at the resonance wavelength by the element being determined. If the resonance wavelength is less than 2800 Å, the intensity of emission from all sources is usually negligible and no problem arises. At longer wavelengths the problem can be decreased but not eliminated by decreasing the slit width.

Fortunately, this type of interference is not a problem in atomic absorption spectroscopy provided that the equipment is modulated. Under these conditions radiation originating from the flame is continuous and is not detected by the detector and readout system. A problem can arise if a mechanical chopper system is used, in which it is possible for the radiation from the flame to fall on the glass surface of the window of the hollow cathode. Here, it can be reflected back through the chopper system down the light path. Under these conditions it will now be chopped and modulated and will appear to originate in the hollow cathode. The problem can be overcome either by using an electrical chopper system or by slanting the face of the hollow cathode window to prevent reflection down the light path.

There is a secondary effect of radiation originating from the flame. If the emission is intense, the photomultiplier detector may become overloaded and noisy. This is particularly likely when highly incandescant flames such as reducing oxyacetylene flames or flames into which benzene has been introduced are used.

Usually radiation interferences are not a problem with atomic absorption spectroscopy unless intensely incandescent flames are used.

3.11 CHEMICAL INTERFERENCE

The process of atomic absorption spectroscopy involves the production of free neutral atoms from the sample. As described earlier, this involves several steps, the last of which is the liberation from the free atoms from a salt residue. This involves breaking the chemical bond through which the sample element is attached to either an inorganic salt or an organic addend. The strength of this bond varies from one compound to another, hence the efficiency of breaking the bond will vary depending on the atomization process and the strength of the bond. This variation is called chemical interference because it depends on the chemical form of the sample element. It effects flame photometry and atomic absorption spectroscopy alike. It is much more of a problem if the atomization process is inefficient, such as when low-temperature flames are used, but it is less of a problem when high-temperature flames are used.

If no measures are taken to correct for this interference, it will result in erroneous analytical data. The problem can be reduced or eliminated in two ways. The first is to use a high-temperature atomizing system such as a oxyacetylene or nitrous oxide-acetylene flame, the second is to use a complexing agent such as EDTA. This is described in Table 3.6. The procedure depends upon the fact that the metal is complexed and so the strength of the bond is constant whatever other chemical entities are present in the sample.

Chemical interference is a major source of error in atomic absorption spectroscopy, but it can be controlled if a suitable atomizer and a complexing agent are used.

3.12 MATRIX EFFECT

It has been observed that the presence of high concentrations of salt (e.g., several percent) results in a decrease in the absorption signal. Typical samples that have this kind of high metal concentration include seawater, urine, and other body fluids. The problem arises in the atomization step. After the solvent has been evaporated, we are left with a residue. If the sample has a high salt content, then the residue will be composed mainly of the salt from the sample. The element being analyzed will be entrapped in the comparatively large quantity of residue formed. As the residue goes through the flame, a lot of energy is required to break it down and liberate the free atoms in order to effect absorption. The efficiency of this process is low, and so it is quite possible for the sample element to pass through the flame remaining in the residue and never be reduced to free atomic state. The net result is a significant loss of sensitivity of the analytical method.

One of the most successful methods for eliminating this problem is simple dilution of the sample, which cuts down the concentration of the salt present. This may not always be possible, particularly if the sample element being determined is already low in concentration. A second successful method is to use solvent extraction, which can be used either to extract the metal of interest or to extract a salt, leaving the metal of interest in an analyzable solution.

One other problem with high-salt-content samples is that they frequently tend to clog the burner, causing a decrease in sample flow rate and a loss of signal. A further problem is that the small residual particles introduced into the flame can scatter the light. To the detector this scattered radiation appears exactly the same as absorbed radiation. However, this can be overcome by dilution of the sample or by using absorption background correction.

REFERENCES

1. M. D. Amos, and J. B. Willis, *Spectrochim. Acta*, 22, 1325 (1966).
2. J. B. Willis, *Nature*, 207, 715 (1965).
3. A. G. Gordon, *The Spectroscopy of Flames*, Wiley, New York, 1957.
4. F. Breck, Fisher Scientific Co., *L.S.U. Symposium on Analytical Chemistry*, 1974.
5. T. Hadeishi and R. D. McLaughlin, *Science*, 174, 404 (1971).
6. K. Fuwa and B. L. Vallee, *Anal. Chem.*, 35, 942 (1963).
7. A. C. Menzies, *Anal. Chem.*, 32, 898 (1960).

Chapter 4

ANALYTICAL APPLICATIONS

The procedures and techniques recommended in this chapter refer to equipment using flame atomizers. Significantly better sensitivity has been achieved using carbon rod and carbon bed atomizers, but these will be considered in Chapter 7.

Atomic absorption spectroscopy has been limited almost exclusively to the determination of metals in liquid samples. Attempts to analyze solid or gas samples have been made and reported in the past, but these have not proved to be very reliable and are not recommended for routine analytical purposes. The normal procedure for analyzing solid and gas samples is to convert them to the liquid state. Solid samples may be dissolved in a suitable acid or other solvent. Gas samples may be passed through a liquid scrubbing agent and the element of interest removed from the gas phase trapped in the liquid phase and subsequently analyzed. It should be noted that any pretreatment step such as that necessary to handle solid or gas samples is a potential source of analytical error, and care should be taken to avoid such problems.

4.1 QUANTITATIVE ANALYTICAL RANGE

Like all other spectroscopic quantitative analytical procedures, there is a maximum and a minimum to the concentration range of application. The minimum of the range is a function of the detection limits of the element under optimum conditions. The ultimate limiting factor is the noise level of the instrument being used. The maximum of the analytical range is determined by the degree of absorption by the sample. At right concentrations the degree of absorption is very high. Small changes in concentration of the sample produce virtually no changes in absorption by the sample. Hence, it is difficult or impossible to measure absorption changes caused by concentration changes in the sample. This is typical of the maximum end of the analytical concentration range.

Extension of Maximum Absorption Range

In practice, the upper concentration limit can be extended by three methods. The first method is simply to dilute the sample until the concentration of the element being determined is reduced to an acceptable analytical range. The second method can be used when elongated burners are employed. By placing the burner perpendicular to the light path rather than along the light path, much of the flame is removed from the optical system. The number of absorbing atoms is decreased and there is reduction in sensitivity. In practice this method is unwieldy, but it can be used when necessary. One problem with this technique is that the calibration curves used for quantitative analysis must be prepared under exactly the same conditions as those under which measurements are taken.

The third method is to use an absorption line with a lower oscillator strength. The degree of absorption is directly related to the oscillator strength of the resonance line used, as can be seen from Eq. (1.4). If an element has two resonance lines, one of which has an oscillator strength ten times as great as the second line, then the calibration curves will cover dynamic ranges one of which is ten times as great as the other (this is illustrated in Fig. 3.2). It is by far the most convenient method of extending the analytical range, because it can be achieved by merely changing the wavelength of the monochromator. No other changes in optical alignment or sample treatment are involved. One difficulty with the method is that not all elements have several resonance lines that can be used at will. In practice, if a solution contains a very high percentage of an element to be determined, then any one of the above procedures can be used and if necessary all three.

4.2 SENSITIVITY LIMITS

The sensitivity limit is that concentration which absorbs 1% of the signal under the conditions used. This is an artificial definition, but for comparative purposes the universal use of the definition has proved an unambiguous measure for reporting sensitivity data. It avoids problems involved in defining noise level and measuring the analytical signal above the noise level. Also, it can be seen that it is independent of the amplification used. These are all meaningful advantages when comparing data from different laboratories.

In practice, the ultimate detection level is determined by the noise level of the signal recorded. For some elements the detection limits are significantly better than those recorded under the definition of sensitivity limits. However, for some elements the reverse is true. This is particularly so when volatile elements such as arsenic and selenium are examined and the pertinent hollow cathodes are noisy.

Most of the data in this chapter refer to sensitivity limits. A summary of the most sensitive sensitivity limits recorded for the various elements in the periodic table are shown in Table 4.1. It should be noted that no data are recorded for the nonmetallic elements although indirect methods for determination of these elements by atomic absorption have been published.

The data show that many elements can be determined at concentrations of one part per million or less using flame atomizers. This sensitivity is perfectly satisfactory for the great majority of the elements analyzed commercially.

For some elements where the noise level is very low, scale expanders have been used to expand the 1–10% absorption range to full scale rather than 10% of the recorder scale. This increases the detection limits by approximately tenfold. In order to use scale expansion it is important that the noise be as low as possible. Most manufacturers of atomic absorption equipment have recommended procedures under which scale expansion can be used with its inherent increase in analytical detectability.

4.3 QUALITATIVE ANALYSIS

Inasmuch as only one element (sometimes up to three) is usually measured at any one time, absorption spectroscopy is very un-

Table 4.1
Primary Wavelengths and Sensitivities for
Atomic Absorption Analysis

Element	Wavelength (Å)	Sensitivity (ppm)	Element	Wavelength (Å)	Sensitivity (ppm)
Al	3092	1.0	Na	5890	0.015
Ag	3281		Nb	4059	20.0
As	1937	1.0	Nd	4634	10.0
Au	2428		Ni	2320	0.1
B	2497	40.0	Os	2909	1.0
Ba	5535	0.4	P	2136	250.
Be	2349	0.024	Pb	2170	0.2
Bi	2231[a]	0.4	Pd	2476	0.5
Ca	4227	0.04	Pr	4951	10.0
Cd	2288	0.03	Pt	2659	1.0
Ce	2761		Rb	7800	0.1
Co	2407	0.15	Re	3460	15.0
Cr	3579	0.1	Rh	3435	0.1
Cs	8521	0.3	Ru	3499	0.5
Cu	3247	0.1	Sb	2176	0.5
Dy	4212	0.7	Sc	3907	0.4
Er	4008	0.9	Se	1960	0.5
Eu	4594	0.6	Si	2516	1.0
Fe	2483		Sm	4297	8.5
Ga	2874	2.5	Sn	2354	0.5
Gd	3684	15.0	Sr	4607	0.01
Ge	2651	2.0	Ta	2715	10.0
Hf	3072	1.2	Tb	4319	5.0
Hg	2537	1.0	Te	2143	0.5
Ho	4054	2.0	Ti	3643	1.0
In	3040	0.7	T1	3776	0.4
Ir	2640	8.0	Tm	4094	0.35
K	7665	0.04	U	3585	50.0
La	3574	45.	V	3184	1.7
Li	6707	0.05	W	2944	1.0
Lu	3081	6.0	Y	4077	2.0
Mg	2852	0.007	Yb	3988	0.1
Mn	2795	0.05	Zn	2139	0.01
Mo	3133	0.5	Zr	3601	10.0

[a]The 3068 Å Bi line is normally considered the most sensitive, but the strength of the adjacent 3067 Å hydroxyl band head renders it ineffective.

suitable for qualitative analysis unless specific elements are being tested for. For an unknown analysis, emission spectrography must usually be resorted to.

4.4 QUANTITATIVE ANALYSIS

Quantitative measurement is one of the ultimate objectives of analytical chemistry. Calibration methods and the reliability of results will be treated in Chapter 5.

4.5 RECOMMENDED PROCEDURES FOR QUANTITATIVE ANALYSIS

Analytical procedures and recommended conditions for obtaining quantitative reliable results are provided by the major atomic absorption instrumentation manufacturers. These conditions very somewhat from one manufacturer to another, but in general there is a significant degree of compatibility among the techniques recommended. The following is a synopsis of conditions recommended for use with flame atomizers. Using these conditions, reproducible quantitative results have been obtained on a routine basis.

Several types of flames have found routine use, including oxyhydrogen, oxyacetylene, air-acetylene, and nitrous oxide-acetylene flames.

The sensitivity is defined as that concentration of solution which will lead to 1% absorption when atomized in a flame. The detection limit is that concentration which provides a signal that is analytically detectable. It is based on the signal-to-noise ratio. The detection limit necessarily depends very much on instrumental conditions—particularly signal damping, which provides a smoother signal. Unfortunately, it is also necessary to take a longer period of time to come to a steady signal. Care must be taken in interpreting detection limits, because frequently the conditions are idealized and may be not be easily achieved in routine operation.

No data from carbon atomizers have been included in the data in this section.

The author wishes to thank Fisher Scientific Company, particularly Dr. Fred Brech, and the Perkin-Elmer Corporation for permission to use their publications on Recommended Procedures. A considerable part of the following information is based on these publications.

ALUMINUM

Absorption wavelength (Å)	Sensitivity (ppm)	Analytical range (ppm)
3093	1.0	5-50
3961	1.3	
3082	1.4	
3944	2.0	
2373	3.3	
2367	4.0	
2575	8.8	

Flame: High reducing (nitrous oxide-acetylene).

Recorded Interferences: Iron and HCl at concentrations greater than 0.2%. H_2SO_4 decreases Al sensitivity, the degree varying with varying with concentration. V effects Al sensitivity, the effect varying with H_2SO_4 concentration. Ti enhances Al sensitivity. No interference from Cu, Ca, Ph, Mg, Zn, Na, PO_4^{-3}, or SO_4^{-2}.

Recorded Sample Types: Steel, cement, lube oil, bauxite, polypropylene, soil, titanium alloys, and uranium.

ANTIMONY

Absorption wavelength (Å)	Sensitivity (ppm)	Analytical range (ppm)
2176	0.5	3-40
2068	0.7	
2311	1.2	

Flame: Air-acetylene (oxidizing).

Recorded Interferences: When using the antimony 2176 Å line, interference can be encountered from the lead 2170 Å line. If lead is present,
the 2311 Å line should be used.

    High concentrations of copper, e.g., 1000 parts per million, ab-
sorbs slightly at the 2176 Å absorption line.

    High acid concentration depresses absorption. Compensation
should be made by also preparing standard solutions with high acid concentrations.

Recorded Sample Types: Gold, urine, and nonferrous alloys.

ARSENIC

Absorption wavelength (Å)	Sensitivity (ppm)	Analytical range (ppm)
1890	1	5-50
1937	2	50
1972	3	

Flame: Hydrogen flames preferable, but air-acetylene (oxidizing) also
usable.

Recorded Interferences: No data.

Recorded Sample Types: Gold.

Analytical Notes: As^0 atoms readily form As_4 and As_2 molecules,
which absorb poorly. The As^0 population is rapidly depleted in flames,
therefore atomic absorption by As^0 is very dependent on flame temperature and position of the light path in the flame.

BARIUM

Absorption wavelength (Å)	Sensitivity (ppm)	Analytical range (ppm)
5535	0.4	3-25
3501		
3071		

Flame: Nitrous oxide-acetylene (reducing).

Recorded Interferences: Ionization interferences should be suppressed by the addition of 1000–2000 $\mu g/ml$ of alkali salt to standards and samples. Chemical interferences occur in lower-temperature flames.

Recorded Sample Types: Alkaline earths, carbonates, lube oils, and petroleum products.

BERYLLIUM

Absorption wavelength (Å)	Sensitivity (ppm)	Analytical range (ppm)
2348	0.024	0.2-4

Flame: Nitrous oxide-acetylene (reducing).

Recorded Interferences: Decreased sensitivity occurs in the presence of large concentrations ($\geqslant 500$ $\mu g/ml$) of aluminum, silicon, and magnesium.

Recorded Sample Types: Urine, various air filters, and paper.

BISMUTH

Absorption wavelength (Å)	Sensitivity (ppm)	Analytical range (ppm)
2231	0.4	2-50
2228	1.5	
3067	2.1	
2062	5.5	

Flame: Air-acetylene (oxidizing).

Recorded Interferences: Dependent on spectral slit width and lamp current.

Recorded Sample Types: Urine.

BORON

Absorption wavelength (Å)	Sensitivity (ppm)	Analytical range (ppm)
2497	40	100-900
2496	100	

Flame: Nitrous oxide-acetylene (reducing).

Recorded Interferences: No data.

Recorded Sample Types: No data.

CADMIUM

Absorption wavelength (Å)	Sensitivity (ppm)	Analytical range (ppm)
2288	0.03	0.2-2.0
3261		

Flame: Air-acetylene (oxidizing).

Recorded Interferences: Little data.

Recorded Sample Types: Urine, blood, tissue, and zirconium alloys.

CALCIUM

Absorption wavelength (Å)	Sensitivity (ppm)	Analytical range (ppm)
4227	0.04	0.2-7
2398	20	

Flame: Nitrous oxide-acetylene (reducing).

Recorded Interferences: Ionization interferences should be suppressed by the addition of alkali salt. Otherwise (unlike in the air-acetylene flame) there appear to be no interferences from chemical suppression. EDTA may be added in 1% solution to remove chemical interference.

Recorded Sample Types: Animal tissue, saliva, blood serum, bone, urine, feces, cement, plants, soil, coal ash, silicates, minerals, lubricating oils, petroleum products, and beer.

Analytical Notes: A sputtering device has been used to study calcium absorption. It was found that 1.0 μg could be detected using the 4227 Å resonance line.

It is claimed that the analytical sensitivity is increased by running the source at higher currents.

Anionic interferences are reduced by using rich air-acetylene flames. In low-temperature flames the sensitivity is reduced and chemical interferences are increased.

The flame profile is quite sharp. Therefore, to get reproducible results it is important to use the same part of the flame for all measurements.

CESIUM

Absorption wavelength (Å)	Sensitivity (ppm)	Analytical range (ppm)
8521	0.3	2.0-15
8943		
4556		
4593		

Flame: Air-acetylene (oxidizing).

Recorded Interferences: Both standards and samples require a high concentration of alkali metal salt to suppress ionization.

Recorded Sample Types: No data.

Analytical Notes: A low-temperature flame such as air-hydrogen may be preferable in order to minimize the ionization effect.

CHROMIUM

Absorption wavelength (Å)	Sensitivity (ppm)	Analytical range (ppm)
3579	0.1	0.5-5
3593	0.1	
4254	0.2	

Flame: Air-acetylene (reducing).

Recorded Interferences: Iron and nickel suppress chromium absorption.

Recorded Sample Types: Lube oils, petroleum products, urine, iron and steel, blood plasma, and feces.

Analytical Notes: Chromium absorption is sensitive to the fuel-to-air ratio. In the case of chemical and/or matrix interferences, the nitrous oxide-acetylene flame is recommended and only slightly reduces the sensitivity.

COBALT

Absorption wavelength (Å)	Sensitivity (ppm)	Analytical range (ppm)
2407	0.15	0.6-5
2425	0.23	
2521	0.38	
2411	0.55	
3527	4.1	
3453	4.2	

Flame: Air-acetylene (oxidizing).

Recorded Interferences: No data.

Recorded Sample Types: Urine, iron and nickel, cement.

COPPER

Absorption wavelength (Å)	Sensitivity (ppm)	Analytical range (ppm)
3247	0.1	0.5-10
3274	0.2	
2178		
2165	0.7	
2226	1.5	
2492	7.0	
2244	16.0	
2442	30.0	

Flame: Air-acetylene (oxidizing).

Recorded Interferences: No data.

Recorded Sample Types: Aluminum alloys, copper alloys, fertilizers, lub oils, mining samples, petroleum products, textiles, plants, soils, biological samples, gold, lead, wine, refractory metals, silicates, and plating solutions.

DYSPROSIUM

Absorption wavelength (Å)	Sensitivity (ppm)	Analytical range (ppm)
4212	0.7	4.0-2.0
4046	0.8	
4187	0.9	
4195	1.0	

Flame: Nitrous oxide-acetylene (reducing).

Recorded Interferences: Ionization interferences require the addition of a large amount of alkali salt ($\geqslant$ 1000 μg/ml) to samples and standards.

Recorded Sample Types: No data.

ERBIUM

Absorption wavelength (Å)	Sensitivity (ppm)	Analytical range (ppm)
4008	0.9	5.0-40
4151	1.3	
3863	1.3	
3893	2.4	
4088	6.0	
3937	6.5	
3810	7.0	
3905	20.0	
3944	21.0	
4607	22.0	

Flame: Nitrous oxide-acetylene (reducing).

Recorded Sample Types: No data.

Recorded Interferences: Ionization interferences require the addition of alkali salt.

EUROPIUM

Absorption wavelength (Å)	Sensitivity (ppm)	Analytical range (ppm)
4594	0.6	3.0-50
4627	0.9	
4662	1.1	
3221	7.0	
3213	9.0	
3111	9.0	
3334	12.0	

Flame: Nitrous oxide-acetylene (reducing).

Recorded Sample Types: No data.

Recorded Interferences: Ionization interferences require the addition of alkali salt.

GADOLINIUM

Absorption wavelength (Å)	Sensitivity (ppm)	Analytical range (ppm)
4079	15	100-1000
3684	16	
3783	16	
4058	17	
4054	18	
3714	25	
4194	40	
3674	43	
4045	49	
3946	100	

Flame: Nitrous oxide-acetylene (reducing).

Recorded Sample Types: No data.

Recorded Interferences: Gadolinium ionizes relatively easily in high-temperature flames, thus reducing sensitivity. Ionization can be controlled by adding between 1000 and 2000 parts per million of alkali salts to both samples and standards.

GALLIUM

Absorption wavelength (Å)	Sensitivity (ppm)	Analytical range (ppm)
2874	2.5	15.0-200
2944 ⎱		
2944 ⎰	2.5	
4172	3.7	
2500	20.1	
2450	21.0	
2720	50.0	

Flame: Air-acetylene (oxidizing). Nitrous oxide-acetylene gives a higher sensitivity (1.0 μg/ml) at 287.4 nm.

Recorded Sample Types: No data.

Recorded Interferences: No data.

GERMANIUM

Absorption wavelength (Å)	Sensitivity (ppm)	Analytical range (ppm)
2651.2	2.0	20-200
2651.6	2.0	
2592	5.0	
2710	6.0	
2755	6.5	
2691	9.0	

Flame: Nitrous oxide-acetylene (reducing).

Recorded Sample Types: No data.

Recorded Interferences: No data.

GOLD

Absorption wavelength (Å)	Sensitivity (ppm)	Analytical range (ppm)
2428	0.3	2.5-20
2676	0.4	
2748	250.0	
3128	240.0	

Flame: Air-acetylene (reducing).

Recorded Interferences: None reported.

Recorded Sample Types: Ores.

HAFNIUM

Absorption wavelength ($\mathring{A}$)	Sensitivity (ppm)	Analytical range (ppm)
2866	1.2	100-500

Flame: Nitrous oxide-acetylene (reducing).

Recorded Sample Types: No data.

Recorded Interferences: Fluoride reduces sensitivity. If fluorides are present in the sample then 0.1% of Hf or ammonium fluoride should be added to both sample and standard.

HOLMIUM

Absorption wavelength ($\mathring{A}$)	Sensitivity (ppm)	Analytical range (ppm)
4104	2.0	
4163	3.0	

Flame: Nitrous oxide-acetylene.

Recorded Sample Types: No data.

Recorded Interferences: Ionization caused by high-temperature flame reduced sensitivity. This can be alleviated by adding large amounts of alkali salt to the sample and standards.

INDIUM

Absorption wavelength ($\mathring{A}$)	Sensitivity (ppm)	Analytical range (ppm)
3039	0.7	
3256	0.7	
4105	2.0	
3511	2.2	
2560	8.5	
2754	20.0	

Flame: Air-acetylene.

Recorded Sample Types: No data.

Recorded Interferences: Slight depression of absorption is caused by a hundredfold excess of aluminum, magnesium, copper, zinc, or phosphate.

IRIDIUM

Absorption wavelength ($\mathring{A}$)	Sensitivity (ppm)	Analytical range (ppm)
2640	8.0	50-1000
7665	8.0	
7699	19.0	
4044	4000.0	
4047		

Flame: Air-acetylene (reducing); air-hydrogen if the 4044 $\mathring{A}$ absorption line is used.

Recorded Sample Types: Noble metals.

Recorded Interferences: Interferences have been encountered from aluminum, gold, calcium, chromium, copper, iron, mercury, hafnium, potassium, lanthanum, magnesium, manganese, sodium, nickel, platinum, palladium, lead, osmium, rhenium, ruthenium, tin, titanium, vanadium, technetium, and zinc. Addition of a thousand parts per million of lanthanum removes interference from sodium and potassium, addition of a thousand parts per million of sodium and 500 parts per million of copper removes interference from aluminum, titanium, nickel, palladium, iron, cobalt, platinum, rhenium, potassium, magnesium, sodium, copper, and less than 500 parts per million of calcium. The addition of 7000 parts per million of copper, 5000 parts per million of sodium removes interference from aluminum, bismuth, calcium, cadmium, copper, chromium, iron, mercury, holmium, potassium, lanthanum, magnesium, manganese, nickel, lead, vanadium, yttrium, or zinc at concentrations up to 1000 parts per million; gold, palladium, platinum, cesium, ruthenium, tellurium, or titanium up to 500 parts per million; and silver up to 20 parts per million. The addition of 1% copper and 1% sodium removes interference from palladium, platinum, rhenium, ruthenium, or osmium.

All additions should be made to both samples and calibration solutions in order for the data to be compatible.

IRON

Absorption wavelength (Å)	Sensitivity (ppm)	Analytical range (ppm)
2483	0.15	1.5
2488	0.2	
2522	0.2	
2719	0.3	
3021	0.4	
2501	0.4	
2167	0.5	
3721	0.7	
2967	0.8	
3860	1.2	
3441	1.6	

Flame: Air-acetylene (reducing).

Recorded Interferences: Nitric acid and nickel reduce sensitivity. Reducing flames control the effect.

Recorded Sample Types: Cement, fertilizer, lube oils, petroleum oils, polyproylene nickel plating solutions, soil, tungsten carbide silicates, gold, seawater, serum, titanium zirconium alloys.

LANTHANUM

Absorption wavelength (Å)	Sensitivity (ppm)	Analytical range (ppm)
5501	45	200-2500
4187	75	
4950	80	
3574	180	
3650	180	
3928	180	

Flame: Air-acetylene (oxidizing).

Recorded Sample Types: No data.

Recorded Interferences: Ionization interferences can be removed by the addition of alkali salts to both sample and standard.

LEAD

Absorption wavelength (Å)	Sensitivity (ppm)	Analytical range (ppm)
2170	0.2	1-10
2833	0.5	2-20
2614	5.0	
3684	12.0	

Flame: Air-acetylene (oxidizing).

Recorded Sample Types: Gasoline, lube oils, petroleum materials, copper alloys, steel, urine, air pollutants, blood, urine, body tissue

LITHIUM

Absorption wavelength (Å)	Sensitivity (ppm)	Analytical range (ppm)
6708	0.03	0.1-3.0
3233	7.0	
6104	100.0	

Flame: Air-acetylene (oxidizing).

Recorded Sample Types: Isotope ratio, natural products.

Recorded Interferences: Ionization interferences can be reduced by the addition of 1000 parts per million of cesium or rubidium to sample and calibration standards.

LUTETIUM

Absorption wavelength (Å)	Sensitivity (ppm)	Analytical range (ppm)
3360	6.0	50-500
3312	11.0	
3377	12	
3568	13	
3989	55	
4519	66	

Flame: Nitrous oxide-acetylene (reducing).

Recorded Sample Types: No data.

Recorded Interferences: No data.

MAGNESIUM

Absorption wavelength (Å)	Sensitivity (ppm)	Analytical range (ppm)
2852	0.007	0.01-0.5
2026	0.15	
2796	0.50	

Flame: Air-acetylene (oxidizing).

Recorded Sample Types: Animal tissue, blood, urine, feces, bone, saliva, soil, plant tissue, aluminum alloys, nickel alloys, cast iron, cement.

Recorded Interferences: No interferences are recorded using the nitrous oxide-acetylene flame. If the air-acetylene flame is used, silicon or aluminum depresses the absorption. The interference can be removed by the addition of 1% lanthanum to both sample and standard.

MANGANESE

Absorption wavelength (Å)	Sensitivity (ppm)	Analytical range (ppm)
2795	0.05	0.5-5.0
2801	0.08	
4031	0.5	
2792	50.0	

Flame: Air-hydrogen (oxidizing).

Recorded Sample Types: Fertilizer, plants, soils, steel, blood plasma, animal tissue, copper alloys, steel.

Recorded Interferences: No data.

MERCURY

Absorption wavelength (Å)	Sensitivity (ppm)	Analytical range (ppm)
2537	1.0	10-300

Flame: Nitrous oxide-acetylene (reducing).

Recorded Sample Types: Urine, body fluids, fish, air pollutants, soil, seawater.

Recorded Interferences: Cobalt absorbs weakly at the 2537 Å resonance lines. Reducing agents reduce mercury to mercury-1 or elemental mercury and increase the sensitivity. This can cause erroneous quantitative analysis.

MOLYBDENUM

Absorption wavelength (Å)	Sensitivity (ppm)	Analytical range (ppm)
3133	0.5	5.0-60.0
3170	0.8	
3798	1.0	
3194	1.0	
3864	1.2	
3903	1.6	
3158	2.0	
3209	4.3	
3112	10.0	

Flame: Nitrous oxide-acetylene (reducing); air-acetylene flames can be used with reduced sensitivity (about 50%) and increased chemical interferences.

Recorded Sample Types: Fertilizer, blood plasma, body tissue, steel, soil.

Recorded Interferences: Calcium. The addition of aluminum or sodium sulfate at a 1000 parts per million decreases the interference.

NEODYMIUM

Absorption wavelength (Å)	Sensitivity (ppm)	Analytical range (ppm)
3349	10	10-700
4634	10	
4924	14	
4719	20	

Flame: Nitrous oxide-acetylene (oxidizing).

Recorded Sample Types: No data.

Recorded Interferences: Ionization interferences can be reduced by the addition of 2000 parts per million of alkali salts to the sample and standard.

NICKEL

Absorption wavelength (Å)	Sensitivity (ppm)	Analytical range (ppm)
2320	0.1	1.0-10.10
2311	0.15	
3525	0.3	
3415	0.4	
3041	0.4	
3412	0.7	
3515	0.8	
3038	1.2	
3370	1.8	
3230	3.0	
2944	5.5	

Flame: Nitrous oxide-acetylene (reducing).

Recorded Sample Types: Petroleum products, urine, steel electroplating solutions, alloys, steel.

Recorded Interferences: Molecular absorption by the flame is common when using the 2320 Å resonance line. No such interference takes place at 3524 Å. In reducing flames the presence of iron cobalt or chromium depresses the absorption. The use of oxidizing flames or a nitrous oxide-acetylene flame removes the interference.

NIOBIUM

Absorption wavelength (Å)	Sensitivity (ppm)	Analytical range (ppm)
3344	20	100-1000
3580	22	
3349	24	
4089	28	
3358	30	
4124	38	
3576	50	
3535	60	
3740	64	
4153	100	

Flame: Nitrous oxide-acetylene (reducing).

Recorded Sample Types: No data.

Recorded Interferences: Ionization interferences can be removed by the addition of 2000 parts per million of alkali salts to the sample and standard.

OSMIUM

Absorption wavelength (Å)	Sensitivity (ppm)	Analytical range (ppm)
2909	1.0	10-200
3059	1.6	
2637	1.8	
3018	3.0	
3302	3.5	
2715	4.0	
2807	4.5	
2644	5.0	
4420	20.0	
4261	30.0	

Flame: Nitrous oxide-acetylene (reducing).

Recorded Sample Types: Noble metals.

Recorded Interferences: No recorded interferences.

PALLADIUM

Absorption wavelength (Å)	Sensitivity (ppm)	Analytical range (ppm)
2448	0.5	5.0-50.0
2476	0.4	
2763	1.3	
3405	1.5	

Flame: Air-acetylene (highly oxidizing).

Recorded Sample Types: Noble metals.

Recorded Interferences: The absorption of palladium is reduced by the presence of aluminum, cobalt, nickel, platinum, rhodium, or ruthenium. The interference can be overcome by the addition of either 0.5% lanthanum or 0.02 molar EDTA to both sample and standard.

PHOSPHORUS

Absorption wavelength (Å)	Sensitivity (ppm)	Analytical range (ppm)
2136 2135	250	2000-10,000
2149	500	

Flame: Nitrous oxide-acetylene (reducing).

Recorded Sample Types: No data.

Recorded Interferences: No recorded interferences.

PLATINUM

Absorption wavelength (Å)	Sensitivity (ppm)	Analytical range (ppm)
2659	1.0	5-100
3065	2.0	
2830	3.5	
2930	3.8	
2734	4.0	
2702	4.5	
2487	5.0	
2998	5.5	
2719	9.0	

Flame: Air-acetylene (oxidizing).

Recorded Sample Types: Noble metals.

Recorded Interferences: The absorption is depressed by high concentrations of lithium, sodium, potassium, magnesium, calcium, copper, lead, strontium, barium, chromium, iron, cobalt, nickel, ruthenium, rubidium, ytterbium, palladium, sulfuric acid, phosphoric acid, perchloric acid, HBr, and ammonium. The interference can be removed by the addition of 2000 parts per million of lanthanum and 0.1 molar hydrochloric acid.

POTASSIUM

Absorption wavelength ($\mathring{A}$)	Sensitivity (ppm)	Analytical range (ppm)
7665	0.04	0.2-2.0
7699	0.08	
4044 ⎱ 4047 ⎰	20.0	

Flame: Air-acetylene (rich)

Recorded Sample Types: Blood serum, urine, saliva, feces, animal tissue, plant tissue, soil, foodstuff.

Recorded Interferences: Potassium is easily ionized in high-temperature flames. This can be reduced by the addition of large excesses of other alkali salts to both sample and standard solutions.

PRASEODYMIUM

Absorption wavelength ($\mathring{A}$)	Sensitivity (ppm)	Analytical range (ppm)
4951	10.0	

Flame: Nitrous oxide-acetylene.

Recorded Sample Types: No data available.

Recorded Interferences: Ionization inteferences can be reduced by the addition of 2000 parts per million of alkali salt to both sample and standard.

RHENIUM

Absorption wavelength (Å)	Sensitivity (ppm)	Analytical range (ppm)
3460	15	100-1000
3465	25	
3452	35	

Flame: Nitrous oxide-acetylene (reducing).

Recorded Sample Types: No data.

Recorded Interferences: No recorded interferences.

RHODIUM

Absorption wavelength (Å)	Sensitivity (ppm)	Analytical range (ppm)
3435	0.1	0.5-20
3692	0.2	
3397	0.3	
3502	0.4	
3658	0.6	
3701	1.0	
3507	4.5	

Flame: Air-acetylene (oxidizing).

Recorded Sample Types: Noble metals.

Recorded Interferences: Numerous elements depress the absorption by rhodium. The interferences are removed by the addition of 1% lanthanum sulfate, 1% copper, and 1% sodium. The addition of 3% sodium sulfate also reduces interferences significantly.

RUBIDIUM

Absorption wavelength (Å)	Sensitivity (ppm)	Analytical range (ppm)
7800	0.1	0.5-7.0
7948	0.2	
4202	12.0	
4216	25.0	

Flame: Air-acetylene (oxidizing); air-hydrogen flames reduces ionization.

Recorded Sample Types: No data.

Recorded Interferences: Ionization interferences can be removed by the addition of 2000 parts per million of another alkali salt to both sample and standard.

RUTHENIUM

Absorption wavelength (Å)	Sensitivity (ppm)	Analytical range (ppm)
2499	0.5	3-50
3728	0.8	
3799	1.1	
3926	6.0	

Flame: Air-acetylene (oxidizing).

Recorded Sample Types: No data.

Recorded Interferences: Molybdenum decreases absorption, platinum or rhodium increases absorption.

SAMARIUM

Absorption wavelength (Å)	Sensitivity (ppm)	Analytical range (ppm)
4297	8.5	50.0-500.0
4760	12.0	
5117	12.0	
5201	14.0	
4728	17.0	
4783	18.0	
4583	22.0	

Flame: Nitrous oxide-acetylene (reducing).

Recorded Sample Types: No data.

Recorded Interferences: Ionization interferences can be reduced by the addition of 2000 parts per million of alkali salt to both sample and standard.

SCANDIUM

Absorption wavelength (Å)	Sensitivity (ppm)	Analytical range (ppm)
3912	0.4	2.0-25.0
3908	0.4	
4024	0.6	
4020	0.7	
4055	1.1	
3270	1.4	
4082˙	3.0	
3274	5.0	

Flame: Nitrous oxide-acetylene (reducing).

Recorded Sample Types: No data.

Recorded Interferences: Ionization interferences can be reduced by the addition of 2000 parts per million of alkali salt to both sample and standard.

SELENIUM

Absorption wavelength (Å)	Sensitivity (ppm)	Analytical range (ppm)
1960	0.5	2.0-50.0
2040	1.5	
2063	6.0	
2075	20.0	

Flame: Air-acetylene (oxidizing); Argon-hydrogen-air flame improves sensitivity to 0.2 parts per million, but chemical interferences are increased. Also, this low-temperature flame is unsuitable for organic solvents.

Recorded Sample Types: Copper, copper alloys.

Recorded Interferences: High concentrations of copper.

SILICON

Absorption wavelength (Å)	Sensitivity (ppm)	Analytical range (ppm)
2516	1.0	5.0-100
2507	2.8	
2528	3.5	
2524	4.0	
2217	4.5	
2211	8.0	

Flame: Nitrous oxide-acetylene (reducing).

Recorded Sample Types: No data.

Recorded Interferences: No recorded interferences.

SILVER

Absorption wavelength (Å)	Sensitivity (ppm)	Analytical range (ppm)
3281	0.1	1.0-20

Flame: Air-acetylene (oxidizing).

Recorded Interferences: None recorded.

Recorded Sample Types: Lube oils, lead sulfide concentrates, urine, gold, aluminum, ores.

SODIUM

Absorption wavelength (Å)	Sensitivity (ppm)	Analytical range (ppm)
5890	0.015	0.1-1.0
5895	0.015	
3302	0.30	1.0-10
3303	0.30	1.0-10

Flame: Air-acetylene (oxidizing). Air-hydrogen gives lower noise, especially at 3302 Å.

Recorded Sample Types: Blood, urine, saliva, body tissue, soil, plant tissue, aluminum alloys, cement, petroleum products.

Recorded Interferences: Ionization interferences take place in air-acetylene flame. These interferences can be reduced by the addition of a 1000 parts per million of other alkali salts to both sample and standard.

STRONTIUM

Absorption wavelength (Å)	Sensitivity (ppm)	Analytical range (ppm)
4607	0.1	1.0-10.0
4078	7.0	

Flame: Air-acetylene (highly reducing). Air-hydrogen: Chemical interferences are reduced in nitrous oxide-acetylene flames, but increased ionization reduces sensitivity. Ionization can be decreased by adding an excess of alkali (2000 ppm).

Recorded Sample Types: Plant tissue, soils, alkali carbonates.

Recorded Interferences: Nitric acid reduces the absorption by strontium. Hydrochloric or sulfuric acid reduces absorption to a lesser extent. Compensation can be made by preparing suitable similar standards.

Interference from silica, aluminum, or phosphorus can be removed by the addition of 1% lanthanum.

Ionization interferences can be removed by the addition of 2000 parts per million of alkali salts to both sample and standard.

SULFUR

Absorption wavelength (Å)	Sensitivity (ppm)	Analytical range (ppm)
1807	9.0	No data

Sulfur was detected by G. F. Kirkbright and M. Marshall [1] using an N_2 purge vacuum monochromator. An electrodeless discharge lamp was the source.

TANTALUM

Absorption wavelength (Å)	Sensitivity (ppm)	Analytical range (ppm)
2715	10.0	600-1200
2608 2609	20.0	
2657	25	
2935	25	
2559	25	
2653	30	
2698	30	
2758	32	

Flame: Nitrous oxide-acetylene (reducing).

Recorded Sample Types: No data.

Recorded Interferences: Fluoride causes increase in absorption. The interferences can be removed by the addition of ammonium fluoride to both sample and standard.

TECHNETIUM

Absorption wavelength (Å)	Sensitivity (ppm)	Analytical range (ppm)
2614 2616	3.0	10-70
2609	12.0	
4297	20.0	
4262	25.0	
3182	30.0	
4238	33	
3636	33	
3173	300	
3466	300	
4032	300	

Flame: Air-acetylene (highly reducing).

Recorded Sample Types: No data.

Recorded Interferences: Calcium, strontium, or barium depresses sensitivity. The interference can be removed by the addition of aluminum. Sulfuric acid causes a severe decrease in sensitivity.

TELLURIUM

Absorption wavelength (Å)	Sensitivity (ppm)	Analytical range (ppm)
2143	0.5	2.0-23

Flame: Air-acetylene (oxidizing); air-hydrogen.

Recorded Sample Types: No data.

Recorded Interferences: No recorded interferences.

TERBIUM

Absorption wavelength (Å)	Sensitivity (ppm)	Analytical range (ppm)
4326·	5.0	40-600
4319	6.0	
3901	8.0	
4062	9.0	
4338	10.0	
4105	20.0	

Flame: Nitrous oxide-acetylene (reducing).

Recorded Sample Types: No data.

Recorded Interferences: Ionization interference can be removed by the addition of 2000 parts per million of alkali salts to both sample and standard.

THALLIUM

Absorption wavelength (Å)	Sensitivity (ppm)	Analytical range (ppm)
2768	0.4	2.0-20.0
3776	1.1	
2380	3.0	
2580	10.0	

Flame: Air-acetylene (oxidizing); Air-hydrogen.

Recorded Sample Types: No data.

Recorded Interferences: No recorded interferences.

THULIUM

Absorption wavelength (Å)	Sensitivity (ppm)	Analytical range (ppm)
3718	0.35	3.0-60.0
4106	0.6	
3744	0.6	
4094	0.7	
4188	0.7	
4204	1.0	
3752	2.0	
4360	3.3	
3410	5.0	

Flame: Nitrous oxide-acetylene (reducing).

Recorded Sample Types: No data.

Recorded Interferences: Ionization interferences can be corrected for by the addition of 2000 parts per million of alkali salts to both sample and standard.

TIN

Absorption wavelength (Å)	Sensitivity (ppm)	Analytical range (ppm)
2246	0.5	5.0-100
2863	0.9	
2355	0.9	
2706	1.5	
3034	1.9	
2547	3.0	
2199	3.2	
3009	3.5	
2335	3.5	
2661	15.0	

Flame: Air-acetylene (highly reducing). Increased sensitivity (threefold) can be achieved by using air-hydrogen or argon-air-hydrogen, but chemical interferences are increased.

Recorded Sample Types: Copper alloys, hydrogen, peroxide stabilizer.

Recorded Interferences: Silica, phosphate, pyrophosphates.

TITANIUM

Absorption wavelength (Å)	Sensitivity (ppm)	Analytical range (ppm)
3653	1.0	6.0-150
3643	1.1	
3200	1.2	
3636	1.2	
3355	1.5	
3753	1.7	
3342	1.7	
3999	1.7	
3990	2.0	

Flame: Nitrous oxide-acetylene (reducing).

Recorded Sample Types: No data.

Recorded Interferences: The presence of high concentrations of iron or aluminum increase sensitivity. If these interferences are present, compensation should be made with the standard and samples used.

Numerous elements increase the sensitivity of titanium, but the addition of 2000 parts per million of KCl removes the interference.

Fluoride enhances titanium absorption. The addition of 0.1 molar ammonium fluoride to the sample and the calibration curve corrects for the interference.

TUNGSTEN

Absorption wavelength (Å)	Sensitivity (ppm)	Analytical range (ppm)
4009	1.0	10-500
2551	1.0	
2944	1.2	
2681	1.4	
2724	1.5	
2947	1.6	
4009	2.0	
2831	2.0	
2896	2.5	
2879	3.5	
4302	8.0	

Flame: Nitrous oxide-acetylene (reducing).

Recorded Sample Types: No data.

Recorded Interferences: No data.

Analytical Notes: Best sensitivity is obtained when an organic solvent such as acetone or alcohol is used.

URANIUM

Absorption wavelength (Å)	Sensitivity (ppm)	Analytical range (ppm)
3515	50	200-2000
3585	50	
3567	80	
3515	150	

Flame: Nitrous oxide-acetylene (reducing).

Recorded Sample Types: No data.

Recorded Interferences: Ionization interferences can be eliminated by the addition of 2000 parts per million of alkali salt to sample and standard.

VANADIUM

Absorption wavelength (Å)	Sensitivity (ppm)	Analytical range (ppm)
3184	1.7	5-150
3066	4.0	
3060	4.0	
3056	5.0	
3202	10.0	
3902	10.0	

Flame: Nitrous oxide-acetylene (reducing).

Recorded Sample Types: Petroleum products.

Recorded Interferences: Al and Ti increase sensitivity; addition of excess Al (100–1000 ppm) to both sample and standard reduces interferences.

YTTERBIUM

Absorption wavelength (Å)	Sensitivity (ppm)	Analytical range (ppm)
3988	0.1	0.7-7.0
3464	0.4	
2464	0.8	
2672	4.0	

Flame: Nitrous oxide-acetylene (reducing).

Recorded Sample Types: No data.

Recorded Interferences: Ionization interferences can be eliminated by the addition of 2000 parts per million of alkali salt to both sample and standard.

YTTRIUM

Absorption wavelength (Å)	Sensitivity (ppm)	Analytical range (ppm)
4102	2.0	20-200
4107	2.5	
4128	2.5	
4143	2.8	
3621	4.0	

Flame: Nitrous oxide-acetylene (reducing).

Recorded Sample Types: No data.

Recorded Interferences: No data.

ZINC

Absorption wavelength (Å)	Sensitivity (ppm)	Analytical range (ppm)
2139	0.01	0.1-1.2
3079	50.0	

Flame: Nitrous oxide-acetylene (reducing).

Recorded Sample Types: Blood serum, urine, animal tissue, plant tissue, soils, rocks, aluminum alloys, copper alloys, fertilizer, electroplating solutions, wines.

Recorded Interferences: High concentrations of silicon.

ZIRCONIUM

Absorption wavelength (Å)	Sensitivity (ppm)	Analytical range (ppm)
3601	10.0	50-600
3548	15.0	
3030	15	
3012	17	
2482	18	
3624	20	

Flame: Nitrous oxide-acetylene (reducing).

Recorded Sample Types: No data.

Recorded Interferences: F^-, NH_4^+, or Cl^- increase sensitivity; SO_4^{2-}, NO_3^-, and $NiBr_2$ decrease sensitivity. Adding 0.1 molar NH_4F to sample and standard removes interference.

REFERENCE

1.   G. F. Kirkbright and M. Marshall, *Anal. Chem.*, 44, 1288 (1972).

Chapter 5

QUANTITATIVE ANALYSIS AND TREATMENT OF DATA

5.1 INTRODUCTION

When a sample is introduced into a flame and absorption by atoms takes place, the data obtained are measurements of the degree of absorption. For quantitative analysis it is necessary to translate these absorption measurements into the concentration of the element being determined in the sample. This is achieved by using a calibration curve or analytical curve which illustrates the relationship between concentration and absorption for the particular element of interest. This is generally done by measuring the absorption of standard solutions of known metal concentration.

It is most important that the solvent used in preparing the standard solution be the same as the solvent of the sample. This will be readily understood if we remember the profound effects that the solvent has on the absorption signal for a given metal concentration.

It is also important to remember that when very dilute solutions are prepared for calibration purposes, the metal can plate out onto the walls of the container within a relatively short period of time (e.g., one or two hours). It is particularly important if the concentration is less than one part per million.

To avoid error caused by plating out of the sample element,

115

it is necessary to prepare standard solutions and run calibration data very shortly thereafter.

To a first approximation, absorption by atoms follows Beer's law. The basic equation is

$$I_1 = I_0 \times 10^{-a,b,c} \tag{5.1}$$

where

$$
\begin{aligned}
a &= \text{absorbitivity} \\
b &= \text{function of the length of the atomizer} \\
c &= \text{concentration of the sample} \\
I_1 &= \text{intensity after passing through the sample} \\
I_0 &= \text{initial intensity}
\end{aligned}
$$

This can be further simplified to

$$A = a, b, c \tag{5.2}$$

where A = absorbance = $\log I_0 / I_1$.

This equation indicates that there is a linear relationship between absorbance and concentration provided that the burner path length is kept constant. This relationship holds only under ideal conditions; it is seldom realized in practice. As concentration increases, its relationship is usually found to deviate from linearity so that a further increase in concentration does not result in a similar increase in absorbance. This deviation can be accommodated by preparing calibration curves. Suitable salts for preparing calibration curves are listed in Table 5.1.

5.2 RELIABILITY OF RESULTS

All analytical results are subject to error. The error may be large or small, reproducible or not reproducible. For analytical data to be meaningful, it is essential that the error involved in the procedure be measurable. The two principal sources of error are called *determinate* and *indeterminate*.

Table 5.1

Salts Suitable for Preparing Calibration Curves

Metal	Salt
Aluminum	$AlCl_3$ (from metal) in approx. 2.25 N HCl
Arsenic	As_2O_3 dissolved in 20% NaOH; neutralize with dilute H_2SO_4
Barium	$BaCl_2$ in water
Bismuth	$Bi(NO_3)_3$ (from metal) in dilute HNO_3
Boron	H_3BO_3 in water
Cadmium	$CdCl_2$ (from metal) in approx. 4.5 N HCl
Calcium	$CaCl_2$ (from $CaCO_3$) in dilute HCl
Cesium	CsCl in water
Chromium	K_2CrO_4 in water
Copper	$Cu(NO_3)_2$ (from metal) in dilute HNO_3
Iron	$Fe(NO_3)_3$ (from metal) in 0.2 N HNO_3
Lead	$Pb(NO_3)_2$ in 1% HNO_3
Lithium	LiCl (from Li_2CO_3) in dilute HCl
Magnesium	$MgCl_2$ (from metal) in dilute HCl
Mercury	$HgCl_2$ in 0.25 N H_2SO_4
Molybdenum	$(NH_4)_2MoO_4$ in 10% NH_4OH
Nickel	$Ni(NO_3)_2$ (from metal) in dilute HNO_3
Potassium	KCl in water
Rubidium	RbCl in water
Silicon	Na_2SiO_3 (from SiO_2) in water
Silver	$AgNO_3$ in water
Sodium	NaCl in water
Strontium	$Sr(NO_3)_2$ in water
Tungsten	$Na_2WO_4 \cdot 2H_2O$ in water (1% NaOH added)
Zinc	$ZnCl_2$ (from metal) in dilute HCl
Zirconium	$ZrOCl_2 \cdot 8H_2O$ in 20% HCl

As mentioned previously, concentration of solutions, particularly very dilute solutions, may change with time. If 1% accuracy is required, it is good practice to prepare standard solutions daily from stock solutions of 500 or 1000 μg/ml. It has been found that the presence of 1% KNO_3 retards the plating of the metal to the sides of the container and lengthens the life of the standard.

Determinate Errors

Determinate error is caused by faults in the analytical procedure or the equipment used. Such errors may be constant errors or they may be proportional to the true answer obtained, in which case they are called *proportional errors.* Sources of determinant error include the following: (a) Operator error may be caused by improper use of equipment or the preparation by the operator of inaccurate calibration data or improper sample handling, and so on. With reasonable care and attention, operator error can generally be reduced to an acceptable level. (b) Another source of determinate error lies with the reagents such as those used for extracting solvents or preparing calibration curves, and so on. These reagents may be improperly labeled, in which case a major error may take place, or they may be contaminated with the element being analyzed, in which case the error may be minor but nevertheless a source of error. (c) The analytical technique may be at fault or may be inapplicable to the sample being analyzed. (d) Error can also arise if the calibration curves have been prepared using a different solvent than that used in the sample.

Indeterminate Errors

After all the determinate errors of an analytical procedure have been detected and eliminated, the method is still not capable of giving absolutely accurate answers. Numerous small errors may be made at each step of the procedure. Errors may be positive or negative, and the total error may be slightly too high or too low. The net error involved is an indeterminate error. All analytical procedures are subject to indeterminate error. The extent of this error can be calculated, and when it is known, the reliability of the method can be stated.

Indeterminate errors may arise from the sensitivity limits of a balance. For example, a balance may be accurate to within 0.001 g, in which case it would not discriminate between two samples that weigh 1.0153 g and 1.0148 g. Each would apparently weigh 1.015 g. Other sources of error include the limit of accuracy of volumetric equipment and the limit of readout of electrical dials or recording instruments. The significance of these effects is that a small error is always involved in analytical determination.

Let us suppose that an analytical procedure has been developed in which there are no determinate errors. If an infinite number of analyses of a single sample were carried out using this procedure, the distribution of results would be shaped like a symmetrical bell. This shape is called *gaussian*. The frequency of occurrence of any given result would be represented graphically by Fig. 5.5.

If only indeterminate errors were involved, the most frequently occurring result would be the true result (i.e., the result at the maximum of the curve). Unfortunately, in practice it is not possible to make an infinite number of analyses of a single sample. At best only a few analyses can be carried out, and frequently only one analysis is possible. We can, however, use our knowledge of statistics to determine how reliable these results should be. The basis of these statistical calculations is outlined below.

Standard Calibration Curves

Calibration curves are determined experimentally by preparing a series of solutions, each with a known concentration of the absorbing solutions. The absorbance of each solution is then measured and a curve relating the experimentally determined absorbance and the concentration of the solution is prepared. The typical calibration curve in Fig. 5.1 was prepared by plotting the absorbance against the concentration of the solution.

When a sample of unknown concentration is to be determined, the absorbance of the solution is measured. From the calibration curve we may determine the concentration of the sample. For example, if A were found to be equal to 0.50, the concentration would be 5.8.

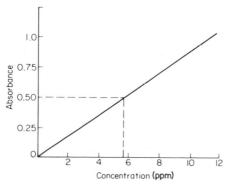

Figure 5.1. Standard calibration curve. From Robinson [1].

A typical series of absorption curves is shown in Fig. 5.2, and the resultant calibration curve is shown in Fig. 5.3.

An experimental determination of the concentration of a solution can be simplified by keeping the intensity of the incident radiation I_0 constant; then only I_1, the intensity of the emerging radiation, needs to be measured. From Eqs. (5.1) and (5.2),

$$A = -\frac{I_1}{I_0} = abc \qquad (5.3)$$

so $-\log (I_1/I_0)$ is related to c by the same curve, and

$$c = -\log \frac{I_1}{I_0} \div ab \qquad (5.4)$$

For a given system, however, a, b, and I_0 are constant. Therefore, a direct relationship exists between c and $-\log I_1$ or $\log (1/I_1)$.

When a solution of unknown concentration is to be determined, it is put into the spectrometer and I_1 (or I_1/I_0) is measured. Using the calibration curve, we ascertain the experimental relationship between I_1 and c. By measuring I_1 when the sample is absorbing radiation, we can determine the concentration of the absorbing compound in the sample.

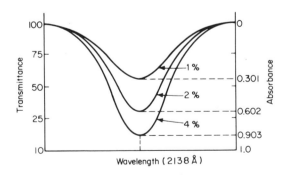

Figure 5.2. Absorption curve for zinc at different concentrations. The line width is greatly exaggerated for the sake of clarity. From Robinson [1].

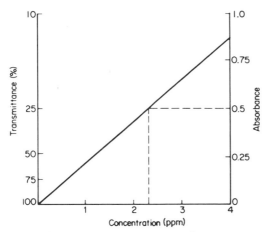

Figure 5.3. Calibration curve for zinc at 2138 Å derived from the absorption curve in Fig. 5.2. From Robinson [1].

Standard Addition Method

The standard addition method may be used if no suitable calibration curves have been prepared and it is undesirable to prepare such a curve, for example, because of the time delay or lack of sufficient information on the solvent in the sample. In a typical example, sodium in a plant stream of unknown composition may be determined by the method of standard addition. Several aliquots, each of 100 ml of the sample, are taken, and different, but known, quantities of the metal being determined are added to each aliquot except one, which is left untreated. The absorption by each aliquot is then measured. The results are shown in Table 5.2.

Table 5.2

Sample	Absorption	Na added to sample (ppm)
1	2.9	0
2	4.2	1
3	5.5	2
4	6.8	3
5[a]	0.5	Background absorption by flame

[a]Flame only; no sample added.

The data can be displayed in the form of a graph (Fig. 5.4). A measure of the background radiation can be made either from a similar sample containing none of the element being determined or by measuring the radiation intensity at a wavelength slightly different from the characteristic wavelength of the metal. It can be seen from Fig. 5.4 that, over the range examined, the relationship between absorption and sodium concentration added was linear. From the difference in readings between samples 1 and 2, 2 and 3, and 3 and 4, it can be seen that 1 ppm of Na caused an increase in absorption of 1.3%. The absorption by the sample with no sodium added was 2.9%. But since the background was 0.5%, the absorption by the sample was $(2.9 - 0.5) = 2.4\%$. But 1.3% is equal to 1 ppm, 2.4% is equal to

$$\frac{1 \times 2.4}{1.3} = 1.85 \text{ ppm}$$

Thus the concentration of sodium in the original sample was 1.85 ppm; it can be calculated from the relationship

$$\text{Concentration} = \frac{R - B}{(\Delta_{em}/\Delta_{conc})} \qquad (5.5)$$

where R is the absorption by the sample, B is the background absorption, and $\Delta_{em}/\Delta_{conc}$ is the rate of change of emission with change in sample concentration.

Similarly, data can be obtained by projecting the Na con-

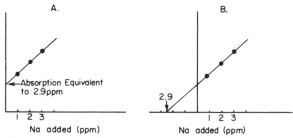

Figure 5.4.   Determination of sodium by the standard addition method.

centration axis in a negative direction (to the left) and measuring the point of intersection of this line and the absorption line. This is shown in Fig. 5.4B.

Definitions for Statistical Calculations

True value, T: the true or real value.

Observed value, V: the value observed by experiment.

Error, E: the difference between the true value T and observed value V (it may be positive or negative):

$$E = V - T \tag{5.6}$$

Mean, M: the arithmetic mean of the observations; that is,

$$M = \frac{\Sigma V}{N} \tag{5.7}$$

where N is the number of observations and ΣV is the sum of all the values V.

Absolute deviation, d: the difference between the observed value V and the arithmetic mean M:

$$d = V - M \tag{5.8}$$

Deviation has no algebraic sign (+ or -); all differences are counted positive. Example 5.1 illustrates some of the statistical values obtained from the three observed values listed under V.

Other definitions include the following.

Relative deviation, D: the absolute deviation d divided by the mean M:

$$D = \frac{d}{M} \tag{5.9}$$

Example 5.1

V	Mean	Error	Deviation	Average deviation, d
103	309/3	0	0	
101		- 2	2	
105	_____	+ 2	2	4/3
309	103	0	4	1.33

Percentage deviation: the deviation times 100 divided by the mean:

$$d(\%) = \frac{(d \times 100)}{M} \qquad (5.10)$$

Relative error: the sum of the errors ΣE divided first by N, then by M:

$$E_{rel} = \frac{\Sigma E/N}{M} \qquad (5.11)$$

In Example 5.1, (a) E_{rel} = d/M = (4/3)/103 = 4/309; (b) d(%) = [4/(3 × 103)] × 100%; (c) E = 0 - 2 + 2 = 0. It can be seen that there are no determinate errors remaining because E = 0. The indeterminate errors, however, give rise to deviation or random differences from the true answer.

Average deviation:

$$d_{av} = \frac{\Sigma d}{N} \qquad (5.12)$$

This is numerically related to the standard deviation, but has no real statistical significance.

Standard deviation:

$$\sigma = \frac{\Sigma(d)^2}{n - 1} \qquad (5.13)$$

Standard Deviation σ. The most commonly used method of presenting the reliability of results is in terms of σ. In Fig. 5.5, the standard deviation coincides with the point of inflection of the curve; 68% of the results obtained fall within $\pm\sigma$ of the true answer and 95% of the results obtained fall within $\pm2\sigma$ of the true answer. These facts enable us to give some meaning to analytical results.

For example, let us suppose that we know by previous testing that the standard deviation σ of a given analytical procedure for the determination of silicon is 0.1%. Also, when we analyze a particular sample using this method, we obtain a result of 28.6% silicon. We can now report with 68% confidence that the true analysis is 28.6 ±0.1%. We could further report with 95% confidence that the true result is 28.6% ±0.2% (where 0.2% = 2σ). Hence, the report that an analysis of a sample indicated 28.6% silicon and that 2σ for the method is 0.2% means that we are 95% confident that the true answer is 28.6% ±0.2% silicon.

Such information concerning the reliability of the method is called precision data. Analytical results published without such data lose much of their meaning. They indicate only the result obtained, not the reliability of the answer.

Variance.   The square of the standard deviation (σ^2) can be used to detect the introduction of error in an analytical method. For example, after a given analytical procedure has been used on numerous occasions to analyze a certain sample, it is possible to obtain the standard deviation for the method.

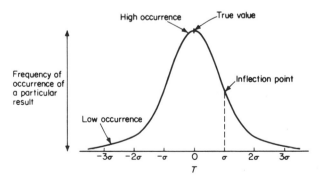

Figure 5.5.   Distribution of results with indeterminate error. The range of the actual error is from low (−) to high (+); T stands for the true value. From Robinson [1].

It might happen that at some time after the analytical method has been developed we suspect that an error has crept into the procedure. We might suspect that a new batch of chemicals used in the process was impure and gave rise to a constant error in the answer. (Such an error would cause a shift in the level of results obtained.) A check on the new results can be made by measuring the variance of the suspect results and comparing it with the variance of the results obtained previously. In this case the variance $(\sigma_2)^2$ of the suspect results is calculated and compared with the variance $(\sigma_1)^2$ of earlier results. The ratio of the variances of these two sets of numbers is called the F function.

$$F = \frac{(\sigma_1)^2}{(\sigma_2)^2} \tag{5.14}$$

Tables of F values can be obtained that indicate the likelihood of of two groups of numbers belonging to the same set or being from two different sets. If the F function shows that the two groups of analytical answers belong to two different sets of numbers, an error has crept into the method and should be eliminated. If the answers all belong to the same set of numbers, no new error has been introduced into the method, and it may be safely used as before.

It should be remembered that the method indicates only the probability of the two sets of numbers belonging to one set. In all cases a judgment has to be made. Of course, when more data are available, the judgment can be reviewed and a more confident decision arrived at.

Precision and Accuracy

It is very important to recognize the difference between precision and accuracy. *Accuracy* is a measure of how close a determination is to the true answer. *Precision* is a measure of how close a set of results are to each other. The difference is illustrated in Table 5.3. A superficial examination of the resulted provided by analyst 2 could be misleading. It is easy to be deceived by the closeness of the answers into believing that his results are accurate. The closeness, however, shows that the results are precise, not that the analysis will result in our obtaining the true answer. The latter must be discovered by an independent method, such as having analyst 2 analyze a solution of known composition (e.g., a solution of pure chemical) and checking his answer against the known composition of the chemical.

It is important to realize that the inability to obtain the correct answer does not necessarily mean that the analyst has poor laboratory technique or is a poor chemist. Many causes contribute to poor analyses, and only by being honest in recording the results obtained can these causes of error be recognized and eliminated.

Table 5.3

Independent Sets of Analytical Results on a Given Sample[a]

	Analyst 1[b]	Analyst 2[c]	Analyst 3[d]
	10.0%	8.1%	13.0%
	10.2%	8.0%	9.2%
	10.0%	8.3%	10.3%
	10.2%	8.2%	11.1%
	10.1%	8.0%	13.1%
	10.1%	8.0%	9.3%
Average	10.1%	8.1%	11.0%
Error	0.0%	2.0%	0.9%

[a]True analysis 10.1% (obtained independently).
[b]Results are precise and accurate.
[c]Results are precise but inaccurate.
[d]Results are imprecise and inaccurate.

Error Analysis

When a new analytical procedure is developed, or when an analytical procedure already in use is put into operation for the first time in a particular analytical laboratory, it is necessary to determine what errors are involved in the method. There are several ways to do this. For example, a sample of known concentration may be analyzed by the new method. If there is a difference between the result obtained by the new analytical method and the calculated result, a determinate error must be present.

Another test for error is to analyze a particular sample by the new procedure and then by a procedure that is known to be free from determinate error. The results obtained by the two methods are then compared. If they agree, the new method is free from determinate error. If they do not agree, an error is involved in the new method.

Each of these methods works well when the error involved is significant and obvious. However, when the difference between the results is small, the question arises, "Is the error caused by a small determinate error or an indeterminate error?" This diffi- culty can be resolved by carrying out a number of analyses by the new method. Using a statistical approach, we can determine if the error is significant and therefore involves a determinate error, or if it is merely part of the distribution of results encoun- tered with indeterminate errors. If a determinate error is involved, it must be eliminated.

Having detected the presence of a determinate error, the next step is to find the source of the error. Practical experience of the analytical method, or first-hand observation of the laboratory operator using the procedure, is invaluable. Much time can be wasted in an office guessing at the source of the trouble. Unex- pected errors can only be discovered in the laboratory. Some of the more common determinate errors are as follows.

Common Determinate Errors

1. Operator Error: Operator errors are caused by the analyst performing the analysis. They may be the result of inexperience; that is, the operator may use the equipment incorrectly, for exam- ple, by placing the sample in the instrument incorrectly each time or setting the instrument to the wrong condition for analysis. Some operator-related errors are (a) carelessness, which is not as common as is generally believed, and (b) poor sampling, which is a common error. If the sample taken to the laboratory is not representative of the complete sample, no amount of good analysis can produce the correct result. Special attention should, therefore, be paid to taking a representative sample for analysis. For example, it is usually fatal to send an untrained assistant to take a sample. It is equally dan- gerous to take the "first cupful" as a good sample. Sample

handling is also important. After a representative sample is received at the laboratory, it should be stored properly. Incorrect storage (e.g., in an open container) can cause contamination by impurities in the air, or evaporation from the sample of its volatile components. After a while the stored material is not representative of the original sample. Another storage problem may be temperature: A sample that is kept too warm may decompose. Remember that even good wine goes sour if it is stored incorrectly. Each of the foregoing errors can be eliminated by proper attention by the operator.

2. Reagents and Equipment: Determinate errors can be caused by faulty reagents. Impurities in the reagents may interfere with the method. The reagents may also be incorrect (e.g., the bottle may be improperly labeled); or the written procedure may call for the wrong compound. These errors can be checked by using a different set of reagents for rechecking the work and by checking to see that the procedure used was the same as that described in the original paper, or by writing to the author.

Numerous errors involving equipment are possible, including incorrect alignment of the instrument, incorrect wavelength settings, incorrect readout values, and incorrect setting of the readout scale (i.e., zero signal should read zero, and for many samples the maximum signal should be set at 100 or some other suitable number). Any variation in these settings can lead to repeated errors. These problems can be removed by a systematic check of the equipment.

3. Analytical Method: The analytical method proposed may be unreliable. It is possible that the original author got good results by compensation of errors; that is, although he appeared to obtain accurate results, his method may have involved errors that balanced each other out. When the method is used in another laboratory, the errors may differ and not compensate each other. A net error in the procedure would result. Errors involved include (a) incomplete reaction for chemical methods; (b) unexpected interference from the solvent or impurities; (c) a high contribution from the blank (the "blank" is the result obtained by going through the analytical procedure, adding all reagents and taking all steps, but adding either no sample or only the pure solvent with no sample); (d) incorrect choice of wavelength for measurement of spectra;

(e) an error in calculation based on incorrect assumptions in the procedure (errors can evolve from an incorrect assignment of formula or molecular weight to the sample); (f) unexpected interferences. Regarding the last, most authors check all the compounds likely to be present to see if they interfere with the method; unlikely interferences may not have been checked. Checking the original article should clear up this point.

There are other sources of error, such as the use of contaminated distilled water in chemical methods, or a change in the line voltage used to operate the instruments. The latter is particularly likely in industrial areas where heavy demands on local power may be made, or released, suddenly. This problem may be acute when the work shift changes or when plants begin operating in the morning or close down for the night. At such times the switching on or off of heavy machinery may significantly change the line voltage available in the laboratory.

By careful checking, all of the foregoing sources of error can be detected and eliminated.

Detection of Indeterminate Error.   Indeterminate errors are always present. The extent of the errors is determined by statistics as described earlier. The standard deviation can be determined accurately only by carrying out an infinite number of tests. This is not a practical proposition. However, an estimate of σ is always greater than true σ to allow for the uncertainty of estimation of σ from a finite number of tests compared to the determination of σ from an infinite number of tests. An increase in the number of observations brings about a decrease in the error in the estimate, so that the calculated value approaches true σ.

A reduction in σ can often be brought about by completely eliminating all determinate errors. If a small error was derived from all the determinate errors mentioned earlier, then the final total error would be random and result in a larger random or indeterminate error. A more skillful operator might reduce any residual determinate error to a small value. The contribution to the standard deviation would be diminished. This is why one operator can carry out an analysis using a particular method and obtain a low standard deviation, whereas another, using the same method and sample, might get results with a higher standard deviation.

In practice the standard deviation for a particular method used for a particular sample may be low when the determination is carried out by a single operator or by the author of the method. If the method is used by ten different operators, however, the standard deviation will be significantly higher, even if the same sample is used each time. These two sets of precision data are called *short-term precision* (obtained by one operator at one time) and *long-term precision* (obtained by several operators at several different times).

Precision can be improved by paying closer attention to the details of the procedure, operating the equipment at peak performance conditions, using correct sampling procedures, and storing the sample properly.

Rejection of Results.   Another source of imprecision is the occasional result that is obviously in error. This result may have been caused by incorrect weighing or measuring, spillage, faulty calculation of results, or carelessness. In any case the result, which includes an error that is neither determinate nor indeterminate, should be rejected and not used in any compilation of results. An acceptable rule to follow is that if the error is greater than 4σ, it should be rejected. When calculating the standard deviation for a new procedure, no suspected result should be included in the calculation of σ. After the calculation, the suspected result should be examined to see if it is greater than 4σ from the true value. If it is outside this limit, it should be ignored; if it is within this limit, the value for σ should be recalculated with this result included in the calculation.

REFERENCE

1. J. W. Robinson, *Undergraduate Instrumental Analysis*, Marcel Dekker, New York, 1970.

Chapter 6

ATOMIC FLUORESCENCE

6.1. INTRODUCTION

The basic process of atomic fluorescence is first the creation of
free atoms from the sample; second, the absorption of radiation
by these free atoms, which become excited in the process; and
third; the reemission of energy as radiation when the atoms return
from the excited state the ground state. Two steps of this process
are similar in many aspects to those observed in atomic absorption
spectroscopy, that is, the creation of free atoms from a sample
and the process of absorption of radiant energy by these atoms.
It is therefore appropriate to devote a section of a book on atomic
absorption to atomic fluorescence, since much of the theoretical
background is common to both fields.

   Atomic fluorescence from excited sodium was first observed
in 1905 by Wood [1] and by numerous other workers since then.
Rigorously studies were initiated and reported, starting with
Winefordner, Vickers, and Staab [2,3] and by a group spearheaded
by T. S. West [4].

   Both research groups indicated that atomic fluorescence holds
several advantages over atomic absorption spectroscopy, particularly the fact that the fluorescence signal is proportional to the intensity of radiation that excites the sample. They project that with
extremely strong excitation radiation, the fluorescence signal will

similarly increase and that there is virtually no theoretical limit
to the intensity of the fluorescence signal that may thus be produced. This projection is in direct contrast to atomic absorption
spectroscopy, where the signal being measured is the *fraction* of
radiation absorbed by the atom population. The absorption signal
is independent of the radiation power, hence it is not increased
by increasing the power of the radiation.

In addition, in atomic absorption spectroscopy it is vital that
the line width of the radiation source be somewhat narrower than
the absorption line width in order to effect maximum absorption
of the radiation signal. In contrast, in atomic fluorescence the
radiation light source does not fall on the detector, and therefore
light sources of any intensity or reasonable line width can be used.
It is particularly helpful if the light source is broad relative to the
absorption band line width, since this ensures maximum absorption
of radiation by the atom population. This results in an increased
population of excited atoms, and therefore an increased fluorescence signal.

6.2 MATHEMATICAL RELATIONSHIPS

Mathematical relationships have been adequately covered by
Winefordner and Vickers [3]. The following is based on their
treatment.

The intensity of fluorescence is proportional to the quantity
of radiant energy absorbed. Therefore,

$$P_F = \phi P_{abs} \tag{6.1}$$

where

P_F = intensity of fluorescence (total)
P_{abs} = quantity of radiant energy absorbed
ϕ = quantum efficiency of the process, that is, the
number of atoms that undergo observed fluorescence transformation per unit time divided by
the number of atoms excited from state 1 per unit
time

The relationship between the absorbed and the incident radiation is given by the expression

$$P_{abs} = P^0 (1 - e^{-k_0 L}) \Delta \nu \quad \text{watts} \tag{6.2}$$

where

$$
\begin{aligned}
P^0 &= \text{intensity of incident radiation} \\
k_0 &= \text{atomic absorption coefficient} \\
L &= \text{length of the absorption cell} \\
\Delta \nu &= \text{half the base width of the absorption profile}
\end{aligned}
$$

If an absorption line is scanned from a point remote from the center of the line, absorption is zero. It increases to a maximum as the center of the line is approached. As the scan leaves the center of the line, the absorption diminishes to zero. The shape of the curve relating absorption and wavelength is the absorption profile of the line.

An approximation to $\Delta \nu$ can be made using the approach made by Willis [5]. If the absorption profile is gaussian in shape, it can be approximated to a triangle. Under these circumstances, we have

$$\Delta \nu = \frac{\pi^{\frac{1}{2}}}{(\ln 2)^{\frac{1}{2}}} \Delta \nu_G \tag{6.3}$$

where $\Delta \nu_G$ is the absorption line spectral width at a half intensity of the gaussian curve. This quantity can be measured. By combining Eqs. (6.1) and (6.2), we have

$$
\begin{aligned}
P_F &= \phi P_{abs} \\
&= \phi P^0 \Delta \nu (1 - e^{-k_0 L}) e^{-k_0 L/2} \cosh \left(\frac{k_0 L}{2}\right)
\end{aligned}
\tag{6.4}
$$

where $\cosh(k_0 L/2)$ is a correction term to accommodate self-absorption by the sample [6].

Equation (6.4) can be rewritten in terms of the intensity I_r by dividing P_F by the area A_F of the cell (a flame in this case) from which radiation is emitted and by 4π steradians. Then

$$I_F = \frac{\phi P_F \, \Delta\nu}{4\pi \, A_F} \, (1 - e^{-k_0 L}) e^{-k_0 L/2} \cosh(\frac{k_0 L}{2}) \tag{6.5}$$

Equation (6.5) is valid only if the frequency of absorption is equal to the frequency of fluorescence, that is, for fluorescence from the first resonance line. If this is not so, the equation is refined to become

$$I_F = \frac{\phi_1}{4\pi \, A_F} \, \phi_2 \frac{\nu_1}{\nu_2} \, P_2^{\,0} \, \Delta\nu_2 (1 - e^{-k_2^{\,0}L}) e^{-k_1^{\,0}L} \cosh(\frac{k_1^{\,0} L}{2}) \tag{6.6}$$

where

ϕ_1 = quantum efficiency

ϕ_2 = number of atoms that reach excited state 1 per unit time, divided by the total number of atoms leaving state 2 per unit time

ν_1 = frequency of fluoresced radiation

ν_2 – frequency of absorbed radiation

$\phi_1 \phi_2 \dfrac{\nu_1}{\nu_2}$ = energy efficiency of the process

$k_1^{\,0}$ = absorption coefficient for fluoresced radiation (which may be self-absorbed)

$k_2^{\,0}$ = absorption coefficient for the absorbed radiation

If several lines fluoresce, the term can be expanded further [3].

As in molecular fluorescence, I_F is proportional to N, the number of absorbing atoms. But as N increases, the term $(1 - e^{-k^0 L})$ approaches 1. Therefore, I_F goes through a maximum as N increases; then quenching begins and I_F decreases.

Where N is small, I_F/N is linear, and

$$I_F = \frac{\phi P^0 \, \Delta\nu \, k^0 L}{4\pi \, A_F} \tag{6.7}$$

But early work by Mitchell and Zamansky [7] derived that

$$k^0 = \frac{(\ln 2)^{1/2} \lambda^2 g_1}{4\pi^{3/2} \, \Delta\nu_D \, g_2} \, N_0 \, A_t \delta(\overline{a\nu}) \tag{6.8}$$

where

$\Delta\nu_D$ = Doppler half-width

g_1/g_2 = a priori statistical weights of atoms in states 1 and 2

A_t = transition probability

λ = wavelength of the absorbing line

$\delta(a\bar{\nu})$ = rate of change of damping coefficient cm^{-1}

By substituting Eq. (6.8) in Eq. (6.7), we have

$$I_F = \frac{(\ln 2)^{1/2}\phi \ \Delta\nu \ L\lambda^2 A_f \ \delta(a\bar{\nu})g_1}{16\pi^{5/2}A_F \ \Delta\nu_D \ g_2} P^0 N$$

$$= CP^0 N \tag{6.9}$$

where C is a constant for any particular experimental arrangement. This leads to the relationship that the intensity of fluorescence is proportional to the incident radiation and the number of atoms that can absorb.

Many of the factors that affect atomic absorption will affect atomic fluorescence. This includes N, the number of atoms in the light path, and f, the oscillator strength of the absorption line. All the factors that affect N in atomic absorption should affect N in atomic fluorescence in a similar fashion, and should include chemical interferences, solvent effects, atomization efficiency, flame composition, and the stability of the neutral atoms. These factors have been discussed in Chapter 3.

In contrast to atomic absorption, the degree of fluorescence of excited atoms should depend on the quantum efficiency of the process. This in turn is dependent on the efficiency of fluorescence, deactivation by other means (for example, collision), or the loss of radiant energy by other processes in the atom.

6.3 ADVANTAGES OF ATOMIC FLUORESCENCE

The attractive features of atomic fluorescence include the following:

1. I_F, the fluorescense intensity, can be increased by increasing P^0, the incident radiation.
2. C (Eq. 6.9) can be increased by increasing L, the size of the flame, and the quantum efficiency.
3. N, the number of fluorescing atoms, is a function of the unexcited neutral atoms in the system. This is inherently higher than the number of thermally excited atoms.
4. A radiation source with wide spectral lines can be used.
5. The intensity of fluorescence is linearly related to the concentration of the sample element over a wide concentration range.
6. The element may be excited at one wavelength and the fluorescence measured at a different wavelength. This eliminates the effect of scattered radiation.

6.4 LIMITATIONS OF ATOMIC FLUORESCENCE

The chief disadvantage of atomic fluorescence is self-absorption of the fluorescence by the sample. This leads to a reversal of the slope of the curve relating I_F, the fluorescence intensity, and the analysis of the sample at high concentrations. It can be overcome by suitable dilution of the sample.

The limitations of atomic absorption also apply. If a flame atomizer is used, metals that form refractory oxides will be difficult to detect; also, with the present development of equipment, only one element can be determined at a time.

The intensity of fluorescence of a particular line can be affected by four types of fluorescence. Of these, sensitized fluorescence in particular can cause a direct interference to the intensity.

6.5 ANALYTICAL PARAMETERS

The Absorption Process

The absorption process is the method by which the atoms become excited. For exactly the same reasons as those considered in atomic absorption, the energy transitions in which the atom is involved concern transition from the highly populated ground state to higher excited states. The degree of absorption depends on the oscillator strength of the absorption line and the number of atoms in the light path. The oscillator strength is a physical property of the particular atomic species and the energy levels concerned. The number of atoms in the light path depends on the atomization process and is subject to the same interferences as those encountered in atomic absorption spectroscopy, particularly chemical interferences.

It should be pointed out that a greater number of atoms will be excited when the intensity of radiation falling on the population is increased. It is therefore very much an advantage to use highly intense light sources. This is in contrast to atomic absorption spectroscopy, where the percentage of light absorbed rather than the total amount of light absorbed is the important factor.

In the excitation process all the empty orbitals with higher energy are available to be filled by the excited electron. In practice, only the first excited state and the second or third excited state are of analytical value. This is because the oscillator strength of transitions between the ground state and higher excited states is comparatively low, and therefore the quantity of radiation absorbed greatly decreases—resulting in a weak fluorescence signal.

The Fluorescence Process

In molecular fluorescence the absorbing and fluorescing species is a molecule. Like molecules, atoms are also capable of being electronically excited and subsequently fluorescing. In contrast, however, molecules are complicated geometric forms that exhibit vibrational and rotational energy levels. Excited molecules may be unstable and lose their electronic energy to other energy forms (such as heat) and not fluoresce. Atoms, however, are the ultimate elemental species and do not contain effective vibrational or rotational energy levels. It is usual, therefore, for most excited atoms to fluoresce. Also, the atomic fluorescence lines are very narrow, since they are not broadened by vibrational or rotational bands.

6.6 TYPES OF FLUORESCENCE [1]

Winefordner and Vickers [3] suggest that there are four principal types of fluorescence. These they define as follows.

Resonance Fluorescence

Resonance fluorescence occurs when the absorption wavelength and the fluorescence wavelength are identical. This transition is illustrated in Fig. 6.1. For all practical purposes it refers to the use of atomic resonance lines associated with the transition between the ground state and the first excited state of the valence electron.

Two difficulties are involved with resonance fluorescence. First, any scattered radiation of the light source by particles in the flame will be at the resonance fluorescence line. Scattered radiation at this wavelength cannot be distinguished from resonance fluorescence. This results in a direct analytical interference giving falsely high results.

A second problem with resonance fluorescence is that frequently the atom may be excited to the higher (e.g., second) excited state, lose energy, and drop down to the first excited state as in stepwise fluorescence (see p. 141). Having reached the first excited state, it fluoresces and descends to the ground state. This fluorescence occurs as a result of a transition between the first excited state and the ground state and is identical in every way to resonance fluorescence. The process results in an enhancement of the resonance fluorescence intensity. It is a direct analytical inter-

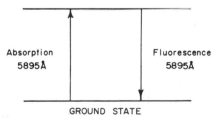

Absorption
5895Å

Fluorescence
5895Å

GROUND STATE

Figure 6.1.   Resonance fluorescence of sodium at 5895 Å.

ference unless careful steps are taken in the calibration procedure
to accommodate this source of radiation.

The principal advantage of resonance fluorescence is that it is
the most intense fluorescence line and provides the highest analytical sensitivity. However, it is also the line most subject to analytical
interferences, and therefore is most difficult to deal with in at-
tempting to get reliable quantitative data.

Direct-Line Fluorescence

Direct-line fluorescence is a process in which the valence
electron is excited from the ground state to a higher excited state.
From this high excited state it falls to a lower excited state; in the
process it fluoresces. The fluorescence is at a lower wavelength
then the absorption wavelength. An illustration is shown in Fig.
6.2. The advantage of using direct-line fluorescence is that it
eliminates interferences encountered with resonance fluorescence,
such as scattered radiation and increased excited atom population.

However, it should be pointed out that the oscillator strength
of the absorption transition between the ground state and the
second excited state is lower than that between the ground state
and the first excited state. This results in fewer excited atoms
available for fluorescence. In addition, only a small fraction of
these will descend to the first excited state prior to fluorescence

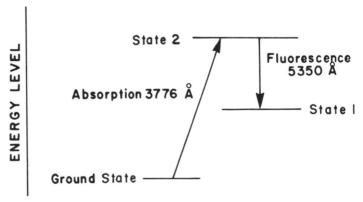

Figure 6.2.   Direct-line fluorescence of thallium.

rather than to the ground state; hence, the process is quite inefficient. It is of reduced anlytical sensitivity but is better for
quantitative determinations. This is particularly so when the sample
concentration is high enough to provide a reasonably intense
fluorescence signal. It should also be pointed out that the transition involved—that is, between the second and first excited state—
is one that is also common to the metal in the hollow cathode, and
it is very likely that this line will be an emitted line from the hollow cathode itself. Unless steps are taken to prevent this line
reaching the atomizer, it can be scattered in the same way as the
resonance line—resulting in a falsely high fluorescence signal. The
problem can be overcome by inserting a filter between the light
source and the atomizer which does not permit radiation of this
wavelength from the hollow cathode to reach the atomizing system.
This in turn prevents scattering of the radiation and eliminates
scatter as a source of error.

Stepwise Fluorescence

In stepwise fluorescence the valence electron is excited to a
higher energy level in the same way as in direct-line fluorescence.

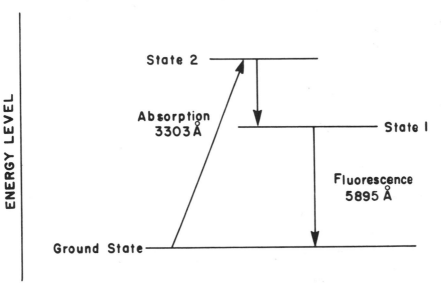

Figure 6.3.   Stepwise fluorescence of sodium.

Here the excited electron loses energy in a nonradiated manner and descends to a lower excited state. Having reached the lower excited state, it then fluoresces and returns to the ground state. The process is illustrated in Fig. 6.3.

Analytically, this fluorescence is of distinctly lower intensity than resonance fluorescence, even though it occurs at the same wavelength as resonance fluorescence. This is because the oscillator strength of the transition between the ground state and the higher excited state is lower than in resonance line absorption, resulting in a decreased number of excited atoms. Of these excited atoms, only a fraction will descend to a lower excited state by a non-radiation process. This results in a relatively low population of atoms in the excited state from which fluorescence occurs. As a result, the analytical sensitivity is significantly lower than that observed in resonance fluorescence. In addition, this line is at the same wavelength as resonance fluorescence, and therefore steps must be taken to prevent scattering of the resonance line which will be strongly emitted by the hollow cathode. As in direct-line fluorescence, this source of interference can be eliminated by inserting a filter between the hollow cathode and the atomizer.

A second analytical interference is an increased population of the final excited state from atoms previously in an upper excited state. This results in an increase in the fluorescence intensity and an apparent increase in analytical sensitivity. Unless care is taken in the calibration procedure, a direct analytical interference will occur.

Thermally Assisted Fluorescence

In thermally assisted fluorescence the excitation procedure is little more complicated than in the previous examples. First, the atom is excited from the ground state to an upper excited state. Here the already excited atom is further excited by thermal collision to an even higher excited state. This process was first observed in 1967 by West and his group [8]. The lines are quite weak, but are of academic interest.

In sensitized fluorescence, excitation first takes place by collision activation to an excited state followed by fluorescence back to the ground state. An example of this process is the behavior of a

mixture of mercury plus thallium vapor. When irradiated at the mercury resonance line at 2537 Å, the mercury atoms become excited.
These excited atoms collide with thallium atoms, which then become excited in the process. The latter fluoresce at Tl 3776 Å and
5350 Å. This is an unusual process and not of practical importance
in atomic fluorescence. It is of course a potential source of interference and should be remembered when setting up quantitative
analytical procedures.

6.7 EQUIPMENT

All the early work in atomic fluorescence used equipment based on
the optical systems and components of flame photometry or atomic
absorption instrumentation. These are illustrated in a schematic
diagram of instrumentation in Fig. 6.4. High-intensity hollow cathode lamps were used as the light source, because they provided
the maximum light power available for exciting the atoms. The
components of the equipment were similar to those used in flame
photometry or atomic absorption. In principle, the sample was
atomized using a flame atomizer, the free atoms were excited using
a hollow cathode lamp, and the fluorescence was measured at right
angles to the excitation source beam. The monochromator was
used to select the pertinent wavelength for measurement.
 Most of the components are the same as those used in

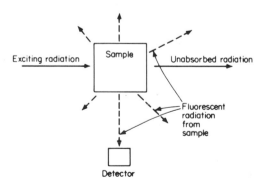

Figure 6.4. Schematic diagram of fluorescence equipment. From Robinson
[9].

atomic absorption spectroscopy and will not be discussed further
in this section. In particular we shall not discuss the atomizer, the
monochromator, and the detector readout system.

Early equipment used line-source radiation sources such as
hollow cathodes. Continuous radiation sources such as hydrogen
lamps have been suggested because all wavelengths are excited
simultaneously. One of the serious disadvantages of continuous
radiation sources is the problem of light scattering. Since the
source emits over a wide wavelength range, any scattering that
takes place in the atomizer (such as the flame) will also take place
over a wide spectral range. Unless steps are taken to eliminate or
correct for this signal, a serious analytical error will result.

Early work on atomic fluorescence was devoted to the use
of continuous sources. However, recently there has been a decrease
in attention to this type of source because of the difficulties involved in getting quantitative analytical data. But its initial potential still remains, in that it is a simple source and has the potential
of exciting many elements in the periodic table.

Radiation Line Sources

The basic premise of the use of line sources is that they emit
radiation at precisely the wavelength that causes excitation of the
element under consideration. Considerable early work was devoted
to the use of metal vapor discharge lamps such as mercury lamps,
hollow cathode lamps, high-intensity hollow cathode lamps, and
demountable water-cooled hollow cathode lamps. These lamps have
been described in earlier chapters on atomic absorption spectroscopy, and their design will not be discussed further.

In general, it can be said that line sources have been quite thor-
oughly studied, and the limit to their potential use has been achieved.
High analytical sensitivity has been obtained, but it should be remembered that these detection limits were generally obtained under
ideal conditions using pure solutions and generally free from interferences. A summary of the results obtained using these line sources
is given in Table 6.1.

It was considered that the most direct way to improve the
analytical detection limits of atomic fluorescence was to increase the

intensity of the radiation source. The electrodeless discharge lamp (EDL) was considered to be a radiation source well worth study. This lamp is described below.

Electrodeless Discharge Lamps

The lamp blank for an electrodeless discharge lamp is successively flushed with argon and evacuated, then charged with suitable metal solvent, recharged with argon at atmospheric pressure or higher, and sealed off. The lamp is then inserted inside a microwave

Table 6.1

Sensitivity (Detection Limits) Obtained in Various Flames with High-Intensity Hollow Cathode Lamps[a]

Element	Line (nm)	Flame	Detection limit (ppm)
Silver	328.1	Air-propane	0.0009
Lead	405.8	Air-hydrogen	0.02
Magnesium	285.2	Air-propane	0.0008
Gold	242.8	Oxygen-argon-hydrogen (sep)	0.005
Iron	248.3	Oxygen-hydrogen-argon	0.02
Iron	b	Air-acetylene	0.08[c]
Copper	324.7	Air-hydrogen	0.001
Cobalt	240.7	Air-hydrogen	0.04
Cobalt	240.7	Oxygen-hydrogen-argon	0.01
Gallium	403.3	Air-hydrogen	0.5
Nickel	232.0	Oxygen-hydrogen-argon	0.003
Nickel	232.0	Air-hydrogen	0.3
Indium	410.5	Air-hydrogen	0.2
Cadmium	228.8	Air-propane	0.0002
Bismuth	d	Air-acetylene (sep)	0.25
Platinum	265.9	Air-hydrogen	50
Palladium	d	Air-acetylene (sep)	2.0
Tin	d	Air-acetylene (sep)	3.0

[a]From West and Cresser [10].
[b]Solar-blind detector.
[c]Estimated value.
[d]Nondispersive technique of Larkins.

discharge cavity. Here it is excited by the microwave radiation and emits at the emission spectrum including the resonance line of the element inside the sealed lamp blank. Schematic diagrams of two electrodeless discharge lamps are shown in Fig. 6.5.

The intensity of these lines is considerably greater than that of the high-intensity hollow cathode lamp. One major problem with this source is its instability. It is vital to quantitative analysis that the lamps be very stable in order to correlate fluorescence intensity with the concentration of the element being determined.

Studies by Winefordner and his group [11] and by West and his group [10] have indicated that sealing the cavity and operating under vacuum provides increased stability.

More recent studies by Windfordner [12] indicate that by heating and maintaining at a constant predetermined temperature the intensity and stability of the lamp can be further improved. Presumably this is because the atoms in the lamp blank are already vaporized by heating the surroundings, and the function of the microwave cavity is merely to cause excitation. It is anticipated that the use of these lamps will lead to further gains in the analytical sensitivity of atomic fluorescence. It is also hoped that the increased stability of the lamps will lead to the achievement of more reliable quantitative data.

Suitable salts from which electrodeless discharge lamps can be made are listed in Table 6.2.

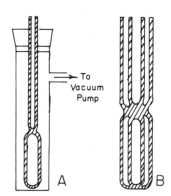

Figure 6.5.   Electrodeless discharge lamps with vacuum jackets. A. Demountable, B. Permanent. From West and Cresser [10].

Modulation of Equipment

In order to avoid error caused by thermal emission from the flame as opposed to fluorescence, it is necessary to modulate the equipment. This can be done by a mechanical chopper or by electronic means. It should also be pointed out that modulation is vital in order to avoid the interferences that would result from the radiation from the flame itself. The thermal emission of the atoms in the flame occurs at exactly the same wavelength as those used in atomic fluorescence, and unless they are corrected for they will result in a direct analytical interference to quantitative analysis. The process of modulation has previously been described (see Chapter 2).

Atomization Processes

As in atomic absorption spectroscopy, a major step in atomic fluorescence is the generation of an atom population from the original sample. As in atomic absorption spectroscopy, the most common atomizer used has been the flame. These systems have been described earlier. Other atomization systems have been developed and successfully demonstrated for atomic fluorescence. These include separated flames, the Massman furnace, and the carbon filament. The latter two will be described in Chapter 7.

Separated Flame Atomizers.   It will be remembered from Chapter 3 that the flame consists of three major sections: the reaction zone, the inner cone, and the outer cone. In the reaction zone the major part of combustion and atom reduction takes place. In the inner cone further reduction and combustion takes place, and here many free atoms exist. In the outer cone air is entrained and combustion is completed, but frequently free atoms are lost.

It is a distinct advantage to be able to extend the inner cone in order to produce a stable population of atoms not disturbed by entrainment of the air. The separated flame, developed by T. S. West, was particularly valuable when used to this end. Earlier designs used quartz tubing placed around the flame, which sepa-

Table 6.2

Electrodeless Discharge Lamps (EDL) for the Excitation
of Atomic Fluorescence (AF)[a,b]

Element	Filler	Notes
Aluminum	$Al\text{-}I_2$	—
Antimony	$Sb\text{-}I_2$; $Sb\text{-}SbI_3$	Cooling often required
Arsenic	$As\text{-}I_2$	—
Barium[c]	$Ba\text{-}Cl$; BaI_2	Not used for AF. Short lived
Beryllium	$Be\text{-}I_2$	—
Bismuth	$Bi\text{-}I_2$; $Bi\text{-}BiI_3$	Optimal sensitivity I_2-EDL
Cadmium	Cd	—
Calcium[c]	CaI_2	Short lived
Chromium	$Cr\text{-}CrI_3$	—
Cobalt	$Co\text{-}CoCl_2$; CoI_2 ; $Co\text{-}CoI_2$; $CoCl_2$	—
Copper[c]	CuI_2 ; $Cu\text{-}I_2$	Short lived
Gallium	$Ga\text{-}I_2$; $Ga\text{-}I_3$; $Ga\text{-}GaI_3$	—
Germanium	$Ge\text{-}I_2$	Cooling required
Gold	$Au\text{-}Cl_2$	Tends to plate out
Hafnium	$Hf\text{-}I_2$	Not used for AF
Iodine	I_2	Used for Bi; poor for I_2
Indium	$In\text{-}I_2$; $In\text{-}InI_3$; InI_3	—
Iron	$FeCl_2$; $Fe\text{-}FeI_2$	$FeCl_2$ more satisfactory
Lead	Pb; PbI_2 ; $Pb\text{-}I_2$	—
Magnesium[c]	MgI_2 ; $Mg\text{-}I_2$	—
Manganese	$Mn\text{-}MnI_2$; MnI_2 ; $MnCl_2$	—
Mercury	Hg; $HgCl_2$	Subject to self-reversal
Molybdenum	$Mo\text{-}MoBr_3$	—
Nickel	$NiCl_2$; $Ni\text{-}NiI_2$	$NiCl_2$ gives Cl_2 background
Palladium	$Pd\text{-}Cl_2$	Not used for AF
Platinum	$Pt\text{-}Cl_2$	Not used for AF
Potassium[c]	K-silicate; K-tetraborate	Not used for AF
Selenium	Se	—
Silicon	$Si\text{-}I_2$	—
Silver	AgI; $Ag\text{-}AgI$; $AgCl$	—
Sodium[c]	Na-silicate; Na-tetraborate; Na-phosphate	Not used for AF
Strontium[c]	$Sr\text{-}SrI_2$	Short lived
Tellurium	Te; $Te\text{-}I_2$	—

Table 6.2 (continued)

Element	Filler	Notes
Thallium	Tl; TlCl; TlI; Tl-I_2	—
Tin	Sn-I_2	—
Titanium	Ti-I_2	—
Vanadium	VCl_3	—
Zinc	Zn	—
Zirconium	Zr-I_2	—

[a]From West and Cresser [10].
[b]EDL emitting line spectra for H, C, N, O, P. S, Cl, Br, He, Ne, Ar, Kr, and Xe have also been prepared though not used in AF. Some of them, e.g., those for S and P, have recently been used in atomic absorption studies.
[c]Denotes rapid chemical attack on silica walls of EDL.

rated the reaction zone from the outer mantel. A later design used an inert gas to separate the flame from the ambient air. A schematic diagram of this burner is shown in Fig. 6.6. An advantage of this system is that the intense radiation from the reaction zone is removed from the optical system. This permits the use of solar-blind detectors and greatly increases the detection limits of atomic fluorescence.

The most common flame is the air-acetylene flame, although the nitrous oxide-acetylene flame has been used extensively, as have oxyhydrogen flames. Argon is the commonly used sheathing gas.

6.8 NONDISPERSIVE ATOMIC FLUORESCENCE SYSTEMS

One of the virtues of atomic fluorescence is that the fluorescence to be monitored can be excited selectively. This is in contrast to flame photometry, where a broad, rich spectrum is emitted by the flame, and to atomic absorption, where a similar rich, intense spectrum is emitted by the hollow cathode discharge lamp. In each of the latter techniques the spectrum falls upon the detector unless special care is taken to select the wavelength of interest using the monochromator.

In contrast, in atomic fluorescence the wavelength of interest can be excited by a suitable radiation source. Fluorescence at the

resonance wavelength occurs and can be measured at right angles
to the excitation beam. If the wavelength of this fluorescence is less
than 3500 Å, then the background emission of the flame is very
low and a nondispersive optical system can be used.

Such a system was greatly improved by the advent of solar-
blind detectors. The latter are insensitive to visible radiation, and
since it is at these wavelengths that most flames are very intense,
the very strong background signal from the flame is not registered
by the detector. The detector, therefore, registers only radiation
at short wavelengths.

A good illustration is the fluorescence of zinc at 2138 Å. Here
the detector can monitor only radiation between 2000 Å and
3000 Å. The zinc may be atomized and excited and will fluoresce
at 2138 Å. The detector registers only the increased signal falling
upon it, which in this case is the fluorescence of the zinc over and
above a very small background of the flame.

Using nondispersive techniques, sensitivity of the method is
greatly increased. Another important feature of the system is the
effort taken to collect as much radiation from the atomizer as
possible. In atomic absorption no special effort is made to do this,
since the only interest is the measure of the degree of absorption
of the hollow cathode lamp, and special collection devices are not
necessary. A schematic diagram of a nondispersive system is shown
in Fig. 6.7.

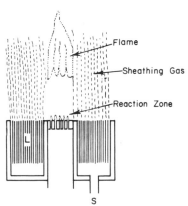

Figure 6.6. Separated flame for atomic fluorescence. L = capillaries for
laminar gas flow; S = sheathing gas inlet. After West and Cresser [10].

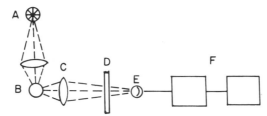

Figure 6.7. Nondispersive atomic fluorescence equipment. A. Light source;
B. Atomizer; C. Gathering lens; D. Filter; E-F. Detector and readout. After
West and Cresser [10].

Multi-Element Analyses

Simultaneous multi-element analyses has been achieved in
atomic fluorescence using the equipment outlined in Fig. 6.8. In
this system a series of light sources, all focused on the atomizers,
are used. Radiation from the atomizer is focused onto a detector
and readout system.

Radiation from the individual light sources falls on the atomizer in sequence. The detectors are programmed to distinguish the
fluorescence generated by each separate radiation source.

This system is comparatively simple to use and quite inexpensive, and offers a convenient way for multi-element analysis.

Since the nondispersive system offers greater development
potential, the components of this system will be described.

Radiation Sources

Two principal types of sources used for atomic fluorescence
are continuous radiation sources and line radiation sources. In the
former, radiation occurs over a wide band, and such sources are
potentially useful for excitation of all the metals in the periodic
table. Radiation line sources are suitable for the excitation of the
particular element used to generate the line.

Continuous Sources: The motivation to use a single continuous
source is that it eliminates the need for many line sources and
should be universal in its application for generation of fluorescence
by different metals.

Hydrogen lamps and deuterium lamps are too weak for practical use in this field. The most successful lamp has been the xenon arc lamp. This lamp is intense and has been used successfully to generate fluorescence from silver, gold, bismuth, calcium, cadmium, copper, chromium, cobalt, iron, indium, manganese, magnesium, nickel, palladium, and zinc.

Using the 150-watt xenon arc lamp, Windfordner et al. [13] detected 13 elements. They used a monochromator and a very sensitive amplifier and detector system. Thompson and West [14] have successfully reported the detection of silver, cadmium, copper, cobalt, iron, magnesium, manganese, lead, and zinc. Interference studies of these elements have not been extensive.

6.9 ANALYTICAL APPLICATIONS

Recent data on detection limits using flame atomizers are given in Section 6.10. The data illustrate the sensitivity of which atomic fluorescence is capable. Care must be taken in translating this into quantitative determinations rather than qualitative detection under ideal conditions.

It should also be noted that the fluorescence signal is very sensitive to changes in operating conditions. In order to obtain good quantitative data, it is necessary that conditions be carefully

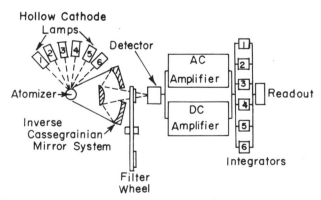

Figure 6.8. Multichannel nondispersive atomic fluorescence spectrometer. After West and Cresser [10].

monitored so that reproducible quantitative data are obtained at all times.

6.10 DETERMINATION OF METALS BY ATOMIC FLUORESCENCE MEASUREMENTS

1. Aluminum: A detection limit of 50 μg/ml at 3961 Å in a premixed nitrous oxide-hydrogen flame has been reported. The fluorescence was attributed to stepwise line fluorescence.
2. Antimony: Fluorescence has been observed for all three resonance lines, 2068, 2176, and 2311 Å. A detection limit of 0.08 μg/ml at 2176 Å in a premixed air-hydrogen flame has been obtained. The source used was an electrodeless discharge lamp.
3. Arsenic: Arsenic fluorescence has been observed at 1890, 1937, and 1972 Å and at 12 other nonresonance lines. Hydrogen flames have been the flames of choice. Optimum sensitivity at low concentrations was obtained at 1890 Å and a detection limit of 0.2 μg/ml was reported using nitrogen-diluted hydrogen-entrained air flame and an electrodeless discharge lamp as the source.
4. Barium: No satisfactory determination reported.
5. Beryllium: Beryllium fluorescence has been observed at 2349 Å. An electrodeless discharge lamp has been the source of choice. The best detection limit of 0.04μg/ml was achieved using a premixed fuel-rich nitrous oxide-acetylene flame on a total-consumption burner.
6. Bismuth: The fluorescence spectrum of bismuth shows many lines, but most investigations have used the 3068 Å line. With an electrodeless discharge lamp as source and a turbulent argon-hydrogen entrained air flame, a detection limit of 0.7 μg/ml was reported at this wavelength.
7. Cadmium: The only analytically useful fluorescence occurs at the 2288 Å resonance line. Electrodeless discharge lamps have yielded the best detection limits. Using an electrodeless discharge lamp as source and turbulent oxyhydrogen flame, a detection limit of 0.00001 μg/ml was achieved.
8. Calcium: Calcium fluorescence has been observed only at the

4227 Å resonance line. Electrodeless discharge lamps have yielded better detection limits than other sources, and a detection limit of 0.02 μg/ml has been reported using a turbulent air-hydrogen flame.

9. Chromium: The most commonly used fluorescence lines for chromium are at 3579, 3593, and 4254 Å. Again, electrodeless discharge lamps as sources have yielded the best detection limits. A detection limit of 0.05 μg/ml has been reported using a laminar air-hydrogen flame and an electrodeless discharge lamp at 3593 Å.

10. Cobalt: Fluorescence has been observed at 2407, 2425, 2521, 2412, 2432, 2414, 2415, and 2529 Å in decreasing order of intensity. Electrodeless discharge lamps have been the best sources, and a detection limit of 0.005 μg/ml has been observed in both air-propane and air-hydrogen flames.

11. Copper: The analytically useful lines of copper are at 3248 and 3274 Å. Air-hydrogen is the flame of choice, and a detection limit of 0.001 μg/ml has been reported using a high-intensity hollow cathode lamp as source. Electrodeless discharge lamps have not been particularly useful for copper due to their relatively short lives.

12. Gallium: The most sensitive lines for gallium are at 4172 and 4033 Å. Hydrogen flames in conjunction with electrodeless discharge lamps have proved most successful. The best detection limit reported was 0.3 μg/ml using an electrodeless discharge lamp as source and an air-hydrogen flame.

13. Germanium: Germanium fluorescence has been observed at 2651 Å in a nitrogen-oxygen-acetylene flame using an electrodeless discharge lamp as source. A detection limit of 15 μg/ml was reported.

14. Gold: Gold fluorescence has been reported at 2428, 2676, 3123, and 3024 Å. The most successful flame employed was the argon-oxygen-hydrogen flame. The best detection limit achieved was 0.005 μg/ml at 2428 Å using a high-intensity hollow cathode lamp as source.

15. Hafnium: No data available.

16. Indium: The most useful fluorescence lines are at 4511 and 4105 Å. Electrodeless discharge lamps are the best sources, and hydro-

gen-oxygen the preferred flame. The best detection limit of 0.1 μg/ml was obtained using the 4105 Å line, an electrodeless discharge lamp as source, and an argon-hydrogen flame.

17. Iron: The iron fluorescence spectrum in flames is complex. The lines at 2483, 2488, 2490, 2491, 2523, and 2527 Å are the most intense, with the 2483 Å line giving the greatest sensitivity. Both hydrogen and acetylene flames have been used equally successfully. The best detection limit reported was 0.008 μg/ml at the 2483 Å line with an air-hydrogen flame and an electrodeless discharge lamp as source.

18. Lead: The most sensitive fluorescence lines for lead occur at 4058, 7229, 2833, and 3640 Å. Hydrogen flames are preferred, and electrodeless discharge lamps have yielded the best detection limits. A detection limit of 0.01 μg/ml at 4058 Å has been reported using a separated argon-oxygen-hydrogen flame and an electrodeless discharge lamp as source.

19. Magnesium: Fluorescence has been observed only at the resonance line at 2852 Å. High-intensity hollow cathode lamps are the preferred sources, and a detection limit of 0.001 μg/ml was reported using this source and an air-acetylene or air-propane flame.

20. Manganese: Manganese fluorescence has been reported at 2795, 2798, 4033, and 4035 Å. Electrodeless discharge sources have been the most successful, and a detection limit of 0.001 μg/ml was recorded using this source and a separated air-acetylene flame.

21. Mercury: To date the only analytically useful line for mercury occurs at 2533 Å. Electrodeless discharge lamps have found the greatest applicability as sources, and a detection limit of 0.08 μg/ml was reported using an air-hydrogen flame.

22. Molybdenum: Fluorescence has been observed at 3133 Å using an electrodeless discharge lamp as source. The detection limit was 0.46 μg/ml using a separated air-acetylene flame.

23. Nickel: The fluorescence spectrum of nickel in flames is complex. The most intense line is at 2320 Å, followed by those at 2311, 3525, and 3415 Å. Electrodeless discharge lamps, high-intensity hollow cathode lamps, and hollow cathode lamps have been equally successful, and little difference has been re-

corded for hydrogen or acetylene flames. A detection limit of 0.003 μg/ml has been reported using various combinations of the above at 2320 Å.

24. Palladium: Palladium fluorescence has been observed at 17 lines, those at 3405, 3635, 3609, and 3243 Å being the most intense. A high-intensity palladium hollow cathode lamp was the most successful source investigated. The best detection limit reported was 0.04 μg/ml using an argon-oxygen-hydrogen flame at the 3405 Å line.

25. Platinum: Platinum fluorescence has been observed at 2659 Å. A detection limit of 50 μg/ml was reported using a shielded hollow cathode lamp as source and an air-hydrogen flame.

26. Potassium: Little data available.

27. Rhodium: Little data available.

28. Ruthenium: Little data available.

29. Scandium: Little data available.

30. Selenium: Selenium fluorescence has been observed at 1961, 2040, 2063, 2075, and 2165 Å. Electrodeless discharge lamps have yielded the best detection limits. Using an air-propane flame and an electrodeless discharge lamp as source, a detection limit of 0.15 μg/ml at 2040 Å has been reported.

31. Silicon: Fluorescence has been observed at 2507, 2514, 2516, 2519, 2524, and 2529 Å. The 2516 Å line is most sensitive. Using an electrodeless discharge lamp as source and nitrogen-separated nitrous oxide-acetylene flame, a detector limit of 0.55 μg/ml was reported.

32. Silver: The analytically useful fluorescence lines occur at 3281 and 3383 Å. Hydrogen flames are preferred, and electrodeless discharge lamps as sources have yielded the best detection limits. A detection limit of 0.0001 μg/ml at 3281 Å was reported using an electrodeless discharge lamp and turbulent air-hydrogen flame.

33. Sodium: Little data available.

34. Strontium: Fluorescence has been reported at 4607 Å. A detection limit of 0.03 μg/ml was obtained using an electrodeless discharge lamp source and an argon-hydrogen entrained air flame.

35. Tellurium: Fluorescence was observed at 2143, 2383, 2386, 2259, and 2531 Å. Electrodeless discharge lamps provide the

best source, and the 2143 Å line has yielded the best detection limits, which was 0.005 μg/ml in an air-propane flame.

36. Thallium: Fluorescence has been observed at 5351, 3776, 3529,
3519, 2768, 2580, and 2538 Å. The most intense line is at
3776 Å. Electrodeless discharge lamps have been the most successfully employed sources, and hydrogen flames are preferred.
A detection limit of 0.008 μg/ml has been reported in an argon-
hydrogen flame.

37. Tin: The fluorescence spectrum shows many lines, the most intense being at 3034 Å. This line is attributed to thermally assisted resonance fluorescence. Electrodeless discharge lamps
have been widely used as excitation sources, and a detection
limit of 0.1 μg/ml was reported at 3034 Å in a nitrogen-separated
nitrogen-oxygen-hydrogen flame.

38. Titanium: Little data available.

39. Vanadium: Vanadium fluorescence has been observed at 2184 Å.
The best detection limit reported was 0.07 μg/ml in air-argon-
separated-nitrous oxide-acetylene flame using an electrodeless
discharge lamp as source.

40. Zinc: The only analytically useful line is at 2139 Å. Electrodeless discharge lamps in conjunction with hydrogen flames have
given the best detection limits. This was 0.00004 μg/ml in a turbulent air-hydrogen flame.

41. Zirconium: Little data available.

6.11 CONCLUSION

It has been amply illustrated that atomic fluorescence is a highly
sensitive method of analytical detection. However, there are a number of difficulties yet to be overcome before many chemists become
convinced that this is a reliable method of quantitative analysis and
can be used on a routine operational basis. Like all comparatively
young techniques, these are teething problems; with time, many of
these difficulties will doubtless be overcome. Atomic fluorescence
will then be able to take its place among the standard methods of
analytical chemistry.

The author wishes to thank Dr. P. J. Slevin for his assistance in
gathering the data on the quantitative applications of atomic
fluorescence.

REFERENCES

1. R. W. Wood, *Phil. Mag.*, 10, 513 (1905).
2. J. D. Winefordner, T. J. Vickers, and R. A. Staab, *Anal. Chem.*, 36, 161 (1964).
3. J. D. Winefordner, *Anal. Chem.*, 36, 165 (1964).
4. T. S. West, *Talanta*, 13, 805 (1966).
5. J. B. Willis, *Australian J. Sci. Res.*, A4, 172 (1951).
6. A. C. Kolb and E. R. Streed, *J. Chem. Phys.*, 26, 1872 (1952).
7. A. C. G. Mitchell and M. W. Zamansky, *Resonance Radiation and Excited Atoms*, Cambridge University Press, New York, 1961.
8. T. S. West, *Talanta*, 14, 1151 (1967).
9. J. W. Robinson, *Undergraduate Instrumental Analysis*, Marcel Dekker, New York, 1970.
10. T. S. West and M. S. Cresser, *Appl. Spectry. Rev.*, 7, 79 (1973).
11. K. E. Zacha, M. P. Bratzel, J. D. Wineforder, and J. M. Mansfield, *Anal. Chem.*, 40, 1733 (1968).
12. R. F. Browner, M. E. Rietta, and J. D. Winefordner, Pittsburgh Conf. on Appl. Spec., paper 136 (March 1972).
13. C. Veillon, J. N. Mansfield, N. L. Pierceson, J. D. Winefordner, and N. L. Kemp, *Anal. Chem.*, 28, 204 (1966).
14. C. Thompson and T. S. West, *Anal. Chim. Acta*, 36, 269 (1966).

Chapter 7

OTHER ATOMIZATION PROCESSES

7.1 INTRODUCTION

The flame atomizer has been used extensively for many years and has proven capable of analyzing a great majority of samples submitted for routine analysis. Much research and development time has been devoted to exploiting the flame atomizer to its limit, and it appears at this time that this atomizer has been developed close to its ultimate potential. By and large, two significant problems have not been successfully solved by the use of flame atomizers. One is the analysis of high solid-containing samples such as seawater, urine, or blood. The second is the analysis of samples available only in limited quantities and in which the interest is to determine trace metal components. Typical examples of the latter are biological clinical samples for which the patient is reluctant or unable to contribute samples of significant size and in which the interest lies in trace metal analysis. A third area of interest is determination of gases, in particular the atmosphere, for trace metal analyses of pollutants.

A common approach to all these analyses had been pretreatment of the sample. This usually took the form of either selective precipitation, selective extraction into a suitable solvent, or—in the case of gas samples—scrubbing out the element of interest with a suitable scrubbing agent. In each case a significant amount of sample was necessary, e.g. several milliliters, which was not always available. Second, the introduction of solvent or precipitant was likely to cause the introduction of impurities, which might contain the element of interest. This in itself represented a serious analytical source of error, and at the low concentration levels of interest was often a very hazardous process.

In the case of air pollution analyses, typical concentrations of metals in air may be 1 $\mu g/m^3$ of air. Under these circumstances, if 1 m^3 of air were scrubbed and the metal extracted completely, then 1 μg of that metal would be available for analysis. This in itself is a challenge, as is scrubbing a cubic meter of air, which may take several hours if it is to be done efficiently. With this background, nonflame methods of analysis were developed, mostly because in the wake of L'vov's work [1] it was recognized that they were capable of very high analytical sensitivity and a true means to translate this into high precision and reliable quantitative analyses.

The flame atomizer is simple to operate (sometimes deceptively simple), and good quantitative data can be achieved even in the hands of a relatively inexperienced operator. The flame atomizer is inefficient in producing atoms and very fast in removing free atoms from the system. Nevertheless, it is quite satisfactory for the analysis of many of the samples brought to the analytical lab. Quantitative analysis at fractional parts per million level is accomplished with relative ease, and in many instances impurity levels below this are of only academic interest. Consequently, commercial equipment invariably uses flame atomizers for atomic absorption spectroscopy. The advent of other systems, however, is of growing importance and doubtless will have a greater impact in the future.

The function of the atomizer is amply described by T. S. West [2] as follows: "The basic requirements of the ideal atom reservoir are an efficient and rapid production of free atoms with the minimum background and background noise, a high of reproducibility and a minimal memory effect a minimal dilution of the atoms. In practice, the development of useful flame and non-flame reservoirs are generally the result of a careful optimization compromise based on these requirements. This development process has very much favored the development and use of flame atomizers. However, recently non-flame atomizers are becoming increasingly important."

7.2 THE L'VOV FURNACE

The first successful nonflame method of analysis developed for
atomic absorption was described in 1961 by L'vov [1]. His system
is shown in the schematic diagram in Fig. 7.1. Basically the instrument is composed of two parts. The first is an electrode onto which
the sample is mounted. The second is a graphite tube heated by
electrical resistance. A conical hole is drilled into the side of the
graphite tube, into which the sample and electrode fit. The latter
are then heated by electrical discharge. Atomization of the sample
takes place inside the graphite tube, and since the latter is maintained at a high temperature, the atomization is quite efficient and
the atoms are maintained for a considerable time in the atomic
state.

In practice, the long tube was heated with a transformer to a
suitable temperature. The inside of the tube was purged with argon

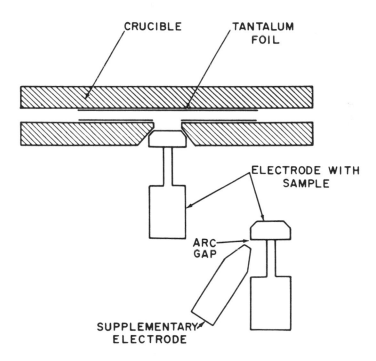

Figure 7.1. High-temperature furnace as designed and used by L'vov.

in order to prevent oxidation of carbon surface. In addition, the surface of the tube, both inside and outside, was treated with pyrographite in order to reduce diffusion through the walls.

After heating up to temperature, which took between 20 and 30 sec, the sample electrode was then inserted in place and the sample vaporized.

In practice, the apparatus was not easy to use, and difficulty was experienced in obtaining reproducible quantitative results. However, the sensitivities obtained by this technique were extremely good. These are listed in Table 7.1. An important contribution of L'vov's work was to encourage further work in flameless atomic absorption research.

In 1960, Walsh [4] developed the cathode lamp as a means of atomization. Here the sample became the cathode and the atoms of the surface were sputtered into the optical light beam. Again very high sensitivity was obtained, but a major difficulty with the technique was that only an analysis of the immediate surface atoms was obtained. This did not necessarily reflect the analysis of the bulk of the sample. Care had to be taken, therefore, in treating the surface in order to be sure that it was representative of the whole sample. In practice, this was difficult to achieve and always held a threat of inaccurate analyses. With many samples, such as metallurgical samples, it is impossible to generate a surface that is truly representative of the sample.

At the Sheffield (England) International Conference on Atomic Absorption Spectroscopy (1969) two papers were presented which described (a) the use of carbon filaments [5] and (b) the carbon bed atomizer [6]. These papers illustrated that highly sensitive reproducible data could be achieved using carbon atomizers. Since that time several other carbon atomizers have been developed and reported upon. These will be described below.

7.3 THE CARBON FILAMENT METHOD

T. S. West and his group at Imperial College London [5] reported the development of the carbon filament method as an atomization process for atomic absorption spectroscopy. The instrument is illus-

Table 7.1

Sensitivity Results (1% Absorption)
with L'vov Furnace[a]

Line (Å)		Measured amount (g)	Sensitivity (g/1% absorption)
Ag	3281	5.0×10^{-13}	1×10^{-13}
Al	3093	2.5×10^{-11}	1×10^{-12}
Au	2428	7.0×10^{-11}	1×10^{-12}
B	2498	5.0×10^{-9}	2×10^{-10}
Ba	5535	1.0×10^{-10}	6×10^{-12}
Be	2349	2.6×10^{-12}	3×10^{-14}
Bi	3068	2.5×10^{-11}	4×10^{-12}
Ca	4227	2.5×10^{-11}	4×10^{-13}
Cd	2288	6.0×10^{-14}	8×10^{-14}
Co	2407	7.5×10^{-12}	2×10^{-12}
Cr	3579	5.0×10^{-11}	2×10^{-12}
Cs	8521	6.6×10^{-12}	4×10^{-12}
Cu	3248	6.3×10^{-12}	6×10^{-13}
Fe	2483	2.5×10^{-11}	1×10^{-12}
Ga	2874	2.5×10^{-11}	1×10^{-12}
Hg	2437	4.5×10^{-10}	8×10^{-11}
In	3039	8.0×10^{-12}	4×10^{-13}
K	4044	0.3×10^{-10}	4×10^{-11}
Li	6708	5.0×10^{-11}	3×10^{-12}
Mg	2852	3.0×10^{-12}	4×10^{-14}
Mn	2795	2.5×10^{-12}	2×10^{-13}
Mo	3135	5.0×10^{-11}	3×10^{-12}
Ni	2320	2.5×10^{-11}	9×10^{-12}
Pb	2833	3.0×10^{-11}	2×10^{-12}
Pd	2476	5.0×10^{-11}	4×10^{-12}
Pt	2659	2.5×10^{-10}	1×10^{-11}
Rb	7800	7.5×10^{-12}	1×10^{-12}
Rh	3435	6.3×10^{-11}	8×10^{-12}
Sb	2311	5.0×10^{-11}	5×10^{-12}
Se	1961	2.0×10^{-10}	9×10^{-12}
Si	2516	2.7×10^{-12}	5×10^{-14}
Sn	2863	1.0×10^{-11}	2×10^{-12}
Sr	4607	2.0×10^{-11}	1×10^{-12}
Te	2143	7.6×10^{-12}	1×10^{-12}
Ti	3653	5.0×10^{-10}	4×10^{-11}
Tl	2768	2.5×10^{-12}	1×10^{-12}
Zn	2138	1.0×10^{-12}	3×10^{-14}

[a]From Robinson and Slevin [3].

trated in Fig. 7.2. Basically the atomizer was a graphite filament approximately 20 mm in length and 2 mm in diameter. The filament
connected two stainless steel electrodes which were water-cooled in
order to prevent them from melting in operation. A very high electrical current, about 100 A at 5 V, was passed through the filament. The latter then became very hot and was capable of thermally
atomizing the sample.

The first approach made to use this instrument was to load
the sample onto the filament, heat it electrically, and measure atomization. It was very quickly realized that this technique would not
work because in an extremely short period of time the solvent and
other inorganic salts were vaporized, causing an extremely high background signal which completely swamped the atomic absorption
signal.

The problem was overcome by West and his group, who developed a very rigorous heating program for the filament.

In practice, a very small sample, e.g., 1 or 2 μl microliters, was
loaded onto the filament. The latter was then heated to a controlled
low temperature for a carefully controlled period of time in order
to evaporate off the solvent. The temperature attained was not sufficiently high to evaporate off the salt residue and cause sample
loss. After a suitable cooling time the filament was then heated
again to vaporize the residue and liberate free atoms. It was found

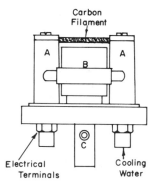

Figure 7.2. Carbon filament atom reservoir. A. Water-cooled electrodes;
B. Laminar flow box; C. Inlet for shield gas. From West and Cresser [2].

Table 7.2

Interferences on Peak Height Measurement
of 1 ppm of Pb[a]

Interferent, 1000 ppm	West-type carbon rod		West-type carbon rod with H_2 flame	
	Pb, 1 ppm	No Pb	Pb, 1 ppm	No Pb
None	100	0	100	0
H_3PO_4	32	19	103	0
NaCl	110	51	92	1
$MgCl_2$	6	11	20	0
$CaCl_2$	0	46	92	1.5

[a]From Amos [7].

that the maximum concentration of free atoms was not at the highest temperature, which existed in the immediate vicinity of the carbon filament, but was some distance away. Very erratic quantitative results were obtained, and the method was highly subject to chemical interference.

The problem was somewhat overcome by sheathing the system and surrounding it with an inert gas or with hydrogen. The same atomization process was adhered to, but the inert gas or hydrogen greatly reduced the chemical interference problems and further increased the sensitivity of the procedure.

An illustration of the improvement is shown in Table 7.2. Other reported sensitivities for this type of atomizer are shown in Table 7.3.

One problem with this type of carbon filament is the formation of carbides by some elements, particularly aluminum, silicon, tungsten, and boron. The problem has been somewhat overcome in the case of aluminum and silicon, but not for tungsten or boron.

A problem with this technique is that the sample soaks into the carbon rod, facilitating the formation of carbides. The use of pyrolytic graphite to coat the surface of the carbon rod has greatly reduced this problem by reducing the porosity of the graphite surface.

Table 7.3

Sensitivity (1% Absorption) and Detection Limit Data
Obtained with Carbon Filament of West[a]

Element	Sensitivity (g/1% absorption)	Detection limit
Ag	6×10^{-10}	2×10^{-9}
Cu	3.3×10^{-11}	5×10^{-11}
Mn	5×10^{-11}	5×10^{-11}
Ni	2.4×10^{-10}	3×10^{-10}
Pb	7×10^{-12}	5×10^{-11}

[a]From Robinson and Slevin [3].

7.4 TANTALUM METAL ATOMIZERS

A tantalum metal analyzer similar in most respects to the carbon
filament analyzer has been developed by J. P. Matousek and B. J.
Stevens [8]. In this system the carbon filament is replaced by a
tantalum metal strip. The heating process and heating program is
the same or similar in character to that used with a carbon filament.
However, the temperature is limited to 2800 °K to prevent the
tantalum metal from melting. The advantage of this device is that
it removes any problem of carbide formation.

Each of these above devices has been used and exploited in
atomic fluorescence spectroscopy. Each is capable of producing
stable atom populations and each therefore meets the prerequisite
of atomizers for fluorescence work.

7.5 THE MASSMAN FURNACE OR HEATED GRAPHITE ATOMIZER

The Massman furnace was developed for commercial use by Perkin-
Elmer Corporation. The schematic diagram is shown in Fig. 7.3.
Basically, the atomizer is a graphite tube 50 mm long and 10 mm
in diameter through which the sample beam passes. A flow of argon
passes through a hollow tube, entering through five small holes and
leaving through the open ends of the tube. The inert gas, usually
argon, flows at a constant rate of about 1.5 liters/min. The entire
tube is maintained inside a water-cooled metal cylinder.

The sample is loaded into the center of the tube. It is then heated by a three-stage electrical heating program. These stages effect drying of the sample to remove the solvent, then ashing of the sample to remove organic material, and finally, a stronger heating step to atomize the sample.

The program must be maintained very rigorously with respect to both the temperature attained at each stage and the duration of each stage. If the ashing temperature is too high, samples can be lost due to sputtering or volatilization. The program varies from sample to sample and element to element. The manufacturers prescribe the program to be used and provide automatic devices that permit selection of the times and temperatures of each stage and automatically run the program of choice. Typical sample sizes are about 20 μl. Reported sensitivities are shown in Table 7.4.

7.6 THE MINI-MASSMAN ATOMIZER

The carbon filament method of West was modified by Varian Associates into the Mini-Massman atomizer. This is illustrated in Fig. 7.4. The atomizer differs from West's filament atomizer as follows: First, the geometry was changed to accommodate a small hole drilled through the rod. Into this hole the sample was placed. In a second modification, the sample was placed on top of the carbon rod. A chimney was placed beneath the rod to allow the introduction of flowing argon, nitrogen, or hydrogen which sur-

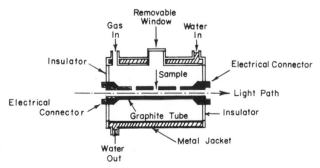

Figure 7.3.   Heated graphite tube atomizer.

Table 7.4

Sensitivities (1%Absorption) in the
Graphite Tube Furnace (HGA-70)

Element	Absolute sensitivity $(g \times 10^{-12})$	20 μl solution $(\mu g/ml)$
Al	150	0.007
As	160	0.008
Be	3.4	0.0002
Bi	280	0.014
Ca	3.1	0.0002
Cd	0.8	0.00004
Co	120	0.006
Cr	18	0.001
Cs	71	0.004
Cu	45	0.002
Ga	1,200	0.06
Hg	15,000	1.5
Mn	7	0.0004
Ni	330	0.016
Pb	23	0.001
Pd	250	0.013
Pt	740	0.04
Rb	41	0.002
Sb	510	0.025
Si	24	0.001
Sn	5,500	0.27
Sr	31	0.0015
Ti	280	0.014
Tl	90	0.0045
V	320	0.016
Zn	2.1	0.0001

[a]From Manning and Fernandez [9].

rounded the site of atomization. The preferred sample size was of the order of 0.5 to 1 μl.

The sample was placed into the hole using a microsyringe. The filament was subjected to a heating program that dried, ashed, and atomized. The program was automatically controlled and could be easily set on the commercial instrument.

Different programs are necessary for the quantitative determination of different elements, since the atomization rate is different for different elements and the evaporation and ashing step also varies from one type of sample to another.

It has been claimed that the use of hydrogen in the cell causes a reducing atmosphere and that in practice the hydrogen ignites when the temperature reaches a suitable level, causing increased sensitivity to be attained.

Analytical sensitivities attained by the Mini-Massman method are shown in Table 7.5.

7.7 THE CARBON BED ATOMIZER HEATED BY RADIOFREQUENCY

Another nonflame atomizer was reported at the Sheffield Conference in 1969 by Robinson [6]. A schematic diagram of the instrument is shown in Fig. 7.5. This instrument was developed originally for the direct determination of metals in air. In practice, the carbon bed was heated up to a temperature of about $1400°C$ with a radiofrequency coil. Air was drawn over the carbon bed and reduction took place according to the equations

$$C + O_2 + N_2 \rightarrow CO + N_2$$
$$MX + CO \rightarrow MO + C + \ldots$$

(7.1)

The technique utilizes the reducing capacities of carbon monoxide and of hot carbon. In contrast to other techniques, there is no direct contact or loading of the sample onto the carbon before heating. On the contrary, the carbon bed is maintained hot. When

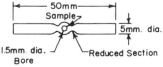

Figure 7.4. Carbon rod (Mini-Massman). Sample is located in the hole in the rod.

in use it is not established that the metal ever comes into direct contact with the hot carbon surface. After atomization the hot gases are drawn into the horizontal T-piece. Free atoms are maintained in the light path while they are in this tube. The optical sample beam passes along this horizontal piece. Atomic absorption takes place at this stage and can be measured in the usual fashion.

The metals lead, mercury, and cadmium have been directly determined in ambient atmosphere. Sensitivities for these metals are shown in Table 7.6.

It has been shown that in addition to air samples (gaseous), liquid samples can also be determined by this technique. The method is quite capable of handling aqueous and nonaqueous samples. A typical sample size is 1 to 2 μl. In the latter case the liquid sample is simply injected onto the top of the hot carbon bed, combustion of the organic compound takes place, and the element of interest is atomized. The combustion products are carbon monoxide and hydrogen for the great majority of organic and aqueous solvents. Carbon monoxide shows very little background absorption, but hydrogen shows significant absorption at wavelengths less than 2300 Å. Under these circumstances a background correction for the several percent absorption is advisable. This can be achieved by using a background corrector (see Chapter 3).

Solid samples have also been analyzed by this technique by simply dropping them onto the carbon bed.

One distinct advantage of this technique is that the solvent is completely burned before it reaches the optical light path. This removes the necessity for a drying and an ashing stage in the procedure. As might be imagined, when volatile elements are being de-

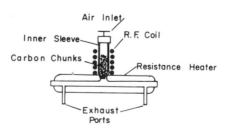

Figure 7.5.   Radiofrequency carbon bed atomization system of J. W. Robinson.

Table 7.5

Sensitivities (1% Absorption) and Detection Limits
for Carbon Rod Atomizer (Mini-Massmann)[a]

Element	Absolute grams	Concentration (μg/ml) for 1 μl sample	Sensitivity (g/1% absorption)
Ag	1×10^{-13}	0.0002	1.2×10^{-12}
Al	3×10^{-11}	0.03	6.3×10^{-11}
As	1×10^{-10}	0.1	9.2×10^{-11}
Au	1×10^{-11}	0.01	2.1×10^{-11}
Be	9×10^{-13}	0.0009	1.1×10^{-12}
Bi	7×10^{-12}	0.007	1.0×10^{-11}
Ca	3×10^{-13}	0.0003	6.5×10^{-13}
Cd	1×10^{-13}	0.0001	6.4×10^{-13}
Co	6×10^{-12}	0.006	1.2×10^{-11}
Cr	5×10^{-12}	0.005	9.2×10^{-12}
Cs	2×10^{-11}	0.02	3.4×10^{-11}
Cu	7×10^{-12}	0.007	2.0×10^{-11}
Eu	1×10^{-10}	0.1	6.1×10^{-11}
Fe	3×10^{-12}	0.003	4.4×10^{-12}
Ga	2×10^{-11}	0.02	2.9×10^{-11}
Hg	1×10^{-10}	0.1	3.4×10^{-10}
K	9×10^{-13}	0.0009	2.0×10^{-12}
Li	5×10^{-12}	0.005	6.1×10^{-12}
Mg	6×10^{-14}	0.00006	4.3×10^{-13}
Mn	5×10^{-13}	0.0005	6.7×10^{-13}
Mo	4×10^{-11}	0.04	4.5×10^{-11}
Na	1×10^{-13}	0.0001	1.4×10^{-13}
Ni	1×10^{-11}	0.01	2.8×10^{-11}
Pb	5×10^{-12}	0.005	6.8×10^{-12}
Pd	2×10^{-10}	0.2	1.5×10^{-10}
Pt	2×10^{-10}	0.2	2.2×10^{-10}
Rb	6×10^{-12}	0.006	4.3×10^{-12}
Sb	3×10^{-11}	0.03	5.3×10^{-11}
Se	1×10^{-10}	0.1	7.2×10^{-11}
Sn	6×10^{-11}	0.06	9.6×10^{-11}
Sr	5×10^{-12}	0.005	6.1×10^{-12}
Tl	3×10^{-12}	0.003	1.1×10^{-11}
V	1×10^{-10}	0.1	9.2×10^{-11}
Zn	8×10^{-14}	0.00008	3.5×10^{-13}

[a]From Robinson and Slevin [3].

Table 7.6

Sensitivities (1% Absorption) Using R.F.-Carbon Atomization[a]

Metal	Sensitivity (g)
Pb	1×10^{-11}
Hg	1×10^{-11}
Cd	2×10^{-13}

[a]From Christian and Robinson [10].

termined, loss of sample can take place during these drying and
ashing stages. This is illustrated in Fig. 7.6.

The elimination of these stages increases the accuracy of the
procedure, a cardinal requirement in analytical chemistry.

Samples that have been analyzed by this method include air,
blood, seawater, urine, and moon dust.

It has also been possible to determine simultaneously the
concentrations of particulate lead and molecular—or "nonfilterable"
—lead in the atmosphere [11]. Other elements that have been
shown to exist in considerable concentrations in the molecular
form as opposed to the particulate form include Hg, Cd, As, Se, and
Cu. At this time the source of these contaminations has not been
identified, but they seem to be caused by evaporation of metal
salts that have very low, but real, vapor pressures.

7.8 THE HOLLOW "T" ATOMIZER

The hollow "T" atomizer was a further development of the radio-
frequency-heated carbon atomizer described immediately above
[12]. It will be remembered that the R.F. atomizer generated a
temperature of approximately 1500°C and that this was the maximum temperature possible because of the use of quartz tubes. In
practice, it was found that the low temperature used was insuffi-
cient to atomize refractory elements reproducibily. The problem
was overcome by using a hollow "T" atomizer made entirely of
carbon and heated electrically.

A schematic diagram of the atomizer is shown in Fig. 7.7.
The atomizer was heated to a temperature of approximately

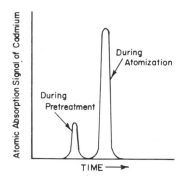

Figure 7.6.   Loss of cadmium during pretreatment.

2600°C by passing an electrical current of 500 A at 12 V.

An inert gas or a slightly oxidizing gas was drawn through the hollow "T" continuously at a prescribed flow rate. A liquid sample was injected into the side arm with a microsyringe. Typical sample sizes were 1 to 2 μl.

Combustion of the sample, including the solvent, was effected in the side stem. Atomization took place during this step. The pro-

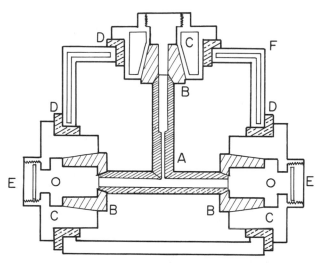

Figure 7.7.   Schematic diagram of hollow "T" atomizer. A. Graphite "T" cell; B. Graphite contacts; C. Water-cooled electrodes; D. Insulators; E. Quartz windows; F. Water-cooled housing. From Robinson and Wolcott [12].

ducts of combustion, usually CO and hydrogen, plus free atoms of the metal under consideration, were then drawn by the flowing inert gas into the crosspiece of the "T". Here the absorption measurements were made.

Typical flow rates were 40 to 50 cm^3/min. The slow flow rate allowed measurements to be taken over a period of several seconds, eliminating the need for fast-response equipment. It was felt that this would contribute to the increased accuracy of the instrument.

Molecular Interference

By burning the solvent in the side stem of the atomizer, the sample fragments and other absorbing species were broken down before reaching the light path. This eliminated the necessity for a drying and ashing step in the analytical process. Some molecular absorption was still encountered from the products of combustion, particularly if the resonance line of the metal being determined was of a short wavelength (e.g., <2400 Å). This was principally caused

Table 7.7

Sensitivity of Measurement with the Hollow "T" Atomizer[a]

Element	Sensitivity (g/0.004 A)
As	1×10^{-10}
Be	1×10^{-10}
Cd	8×10^{-14}
Cr	8×10^{-10}
Cu	1×10^{-10}
Hg	6×10^{-10}
Mn	2×10^{-11}
Ni	4×10^{-10}
Pb	2×10^{-13}
Se	4×10^{-10}
Zn	b

[a]From Robinson and Wolcott [12].
[b]The sensitivity of zinc could not be determined because all the samples of "pure" water obtained gave 100% absorption, and standard solutions could not be prepared.

by the presence in the combustion products of H_2 or CO_2, which absorb at the shorter wavelengths. A background corrector was necessary to eliminate this source of error.

Chemical Interferences

It has been shown that with the carbon filament considerable interferences were encountered in the determination of manganese [13]. The chemical interferences encountered were attributed to the varying rates of atomization of different chemical forms of the manganese. Inasmuch as the atomization process is very rapid, the rate of reaction directly affects the number of free atoms formed at any particular moment. Essentially this was a chemical interference, and directly influenced the atomic absorption measurements observed. In order to compare the hollow "T" system with the carbon filament system, a similar study was carried out on manganese. Results showed a high degree of freedom from interference.

Sensitivity

The results of a preliminary study of the sensitivities that can be achieved with the hollow "T" atomizer are shown in Table 7.7. It can be seen that the sensitivities achieved are at least as good as other atomizers presently available and better than most. It can also be seen that metals that form refractory oxides, such as beryllium and chromium, are atomized quite easily. These preliminary data on a few elements indicate that the atomizer has the potential of being used for the determination of most metallic elements found in the periodic table. The sensitivity appears to be good, background interference seems to be very low, and chemical interference also seems to be much reduced. It has also been demonstrated that the atomizer is capable of analyzing solid, liquid, or gas samples without prior treatment.

This high sensitivity should permit the determination of part-per-billion concentrations in microliter samples and should find application in biomedical samples. The method does not require prior sample treatment of any kind, and should therefore not be susceptible to many errors of contamination or preheat losses encountered with other techniques.

7.9 PRACTICAL APPLICATIONS OF NONFLAME ATOMIC ABSORPTION SPECTROSCOPY

The principal advantage of nonflame absorption spectroscopy is the high sensitivity of the method. It is quite capable of handling very small samples—of the order of microliters—and usually does not require pretreatment of a sample with accompanying introduction of analytical errors. Typical applications of the Mini-Massman atomizers are the determination of magnesium, iron, copper, lead, and zinc in blood and blood plasma; and of silver, aluminum, copper, chromium, magnesium, nickel, and lead in lubricating oils. With the heated graphite Massman atomizer, practical applications include the determination of copper and strontium in milk. Using the tantalum filament, aluminum, arsenic, cadmium, cobalt, chromium, copper, iron, manganese, nickel, lead, tin, and zinc have been determined in natural waters, and lead and chromium have been determined in blood and plasma samples. In addition, copper and manganese have been analyzed in brain tissue by this method.

7.10 CONCLUSIONS

Nonflame atomization techniques have shown themselves capable of generating very high analytical sensitivities. One of their most attractive features is the ability to handle very small sample sizes, of the order of microliters. This is particularly advantageous in clinical analyses.

For some of the techniques the problems of molecular absorption and matrix interferences are still very great, but these are much less of a problem when the R.F. carbon bed atomizer is used.

The precision of the technique is typically between 10 and 20% relative standard deviation. This may seem to be a relatively high level of imprecision, but at concentrations of 10^{-11} and 10^{-12} g it represents an extremely small quantity of material. Frequently, at these concentration level analyses accurate to one significant figure are sufficient to provide the information sought.

At the present time flame atomizers still handle the bulk of samples most successfully. However, it must be conceded that nonflame methods are making great strides. They have already demon-

strated their capability of producing high analytical sensitivity, and
with improved atomization processes may challenge the flame as
the principal atomization process used in routine analysis.

REFERENCES

1. B. V. L'vov, *Spectrochim. Acta*, 17, 761 (1961).
2. T. S. West and M. S. Cresser, *Appl. Spec. Rev.*, 7, 107 (1973).
3. J. W. Robinson and P. J. Slevin, *Amer. Lab.*, p. 10 (August 1971).
4. B. M. Gatehouse and A. Walsh, *Spectrochim. Acta*, 16, 602 (1960).
5. T. S. West, Sheffield International Conference on Atomic Absorption Spectroscopy, 1969.
6. J. W. Robinson, Sheffield International Conference on Atomic Absorption Spectroscopy, 1969.
7. M. D. Amos, *Amer. Lab.*, p. 57 (August 1972).
8. J. P. Matousek and B. J. Stevens, *Anal. Chem.*, 17, 363 (1971).
9. D. C. Manning and F. Fernandez, *Atomic Absorption Newsletter*, 9, 65 (1970).
10. C. M. Christian III and J. W. Robinson, *Anal. Chim. Acta*, 56, 466 (1971).
11. J. W. Robinson and D. K. Wolcott, *Environmental Lett.*, 6, no. 4, 321 (1974).
12. J. W. Robinson and D. K. Wolcott, *Anal. Chim. Acta*, in press.
13. L. Edbow, G. F. Kirkbright, and T. S. West, *Anal. Chim. Acta*, 58, 39 (1972).

INDEX

A

Absorption, coefficient, 7, 8, 134
  frequency, 135
  line width, 20-22, 65, 133
  profile, 134
  signal, 3, 10, 38, 41, 50-52, 59, 60,
    68-70, 75, 115, 133
  wavelength, 2, 10
Air pollution, 159-160
Aluminum, 53, 59, 84, 117, 153
Amos, 57-58
Amplifiers, 42-43
Anions, 59-61
Antimony, 84, 153
Argon, 22, 24, 145, 162, 166
Arsenic, 81, 85, 117, 153
Aspiration, 52, 60
  rate of, 38
Atom cloud, 22-23
Atomic fluorescence, 132-158
  advantages of, 136-137
  limitations of, 137
Atomization, efficiency of, 3-4, 11, 13,
  36-37, 49, 53, 136, 161
  factors affecting, 51-57
    droplet size, 38, 52-54
Atomizers, 19, 35-40
  temperature, 3-10

B

Barium, 86, 117, 153
Barnes demountable hollow cathode,
  23, 27-28
Beer-Lambert law, 8-10, 116
Beryllium, 86, 117, 145, 153
Bismuth, 86, 117, 145, 153
Blood, 159, 176
  plasma, 176
Boron, 87, 117
Bowling Burner, 40
  design, 53-54
  laminar flow, 58
  Lundegardh, 38-40, 50, 52-54
  mechanical, 69
  total consumption, 37-38, 50, 52,
    70-71

C

Cadmium, 87, 117, 145, 152, 153,
  173
Calcium, 87, 117, 153-154
Calibration curves, 9, 71, 80, 115-116,
  119-120
Carbon bed atomizer, 162
Carbon filament atomizer, 11, 12, 37,
  79, 84, 162-165
Carbon tetrachloride, 71